WINGED MIGRATION

WINGED MIGRATION

JACQUES PERRIN

TEXT BY

JEAN-FRANÇOIS MONGIBEAUX

PREFACES BY

JACQUES PERRIN AND JEAN DORST

GALATÉE FILMS PHOTOGRAPHERS

MATHIEU SIMONET

RENAUD DENGREVILLE

GUILLAUME POYET

RENAN MARZIN

SEUIL CHRONICLE

T he late Jean Dorst liked to describe himself as a naturalist. An heir of the Enlightenment, his joy was to discover and his happiness to share. He taught only the laws of nature, gave his science a voice, and rendered it accessible to all by turning it into a wonderful tale.

To understand the migrations of birds, his reasoning took poetic paths, because for him to scientifically elucidate a mystery, to solve an enigma, was also to risk partially effacing a dream.

He could be bewildered by those who seek to know all without having carefully observed beforehand. His precept might have been, "First go out and look, listen, and try to understand, then perhaps I'll tell you more." Only emotion led him to understanding.

When he kept watch over the night sky in Paris, it was because he knew that that night, above the city tumult, a flock of cranes or geese would be crossing it on their way to distant climes. Others also saw them, but he knew where they were going.

CONTENTS

igh over soaring peaks, birds journey through the night sky, an arrowhead of hope fleeting toward the horizon. Only in our imagination can we join them, fly with them, in league with the heavens, guided by the stars.

On they streak through the blue yonder, battling against winds, braving storms, glacial temperatures, and scorching heat, undeterred by confusing fogs and clouds. Driven by an inner certainty, beckoned by the call of the seasons, they know there can be no turning back.

There are birds with which we fly in our dreams, and there are those that show us the way, that serve as examples of the never-ending struggle for survival, landing on a branch, appearing for only an instant before flying away again into a hostile world of predators and redoubtable weather conditions.

When we see a bird flitting through the country sky, skimming over meadows and flowering valleys, we do not realize just what a feat it has already accomplished, that it has already come thousands of miles to reach us, and that those same thousands of miles separate it from its wintering grounds in distant latitudes, beyond vast stretches of ocean and reputedly uncrossable deserts. So, how does it do this?

For animals, the hereditary transmission of knowledge is an innate gift. A few weeks after hatching, birds know how to feed themselves, play, walk, fly, and avoid danger. In a few months, they know how to follow an aerial migration route that has never been shown to them. They know what we perhaps once knew instinctively, the operators forgot, and are now trying to relearn. They know the secrets of the seasons, magnetic poles, stars, and tides from birth. In the egg, they already know the entry codes to the great secrets of the machinery of our universe. And, in no time at all, they know how to reach the most distant climes.

Stuck here on the ground, watching birds pass us by overhead, we started work on this film. To get close to them we would have to go far and high, reaching for the stars.

How could we do this? Man has dreamed of birds from time immemorial. "How could we possibly be the very first to transform this dream into reality?" I asked myself. But we did, and I will always remember the first time we succeeded. A camera operator was following a flight of barnacle geese, while an assistant camera operator was shooing away geese that came too close to the camera with one hand and pulling focus with the other. Before we knew it, we had shot the whole reel of film. Radiant, tears in their eyes, the operators looked at me without a word. What they had just filmed had suddenly rendered their technical mastery meaningless. All that mattered for them was that birds in flight had taken them into their confidence.

JACQUES PERRIN

Bar-headed geese in the Far East.

A hymn to life

One autumn evening, as I was crossing the Jardin des Plantes in the center of Paris, lost in thought, I didn't notice at first other late walkers all gazing skyward.

High above us in the hazy evening blue, a flight of cranes was soaring over the city. Watching these intrepid travelers winging their way toward their winter quarters, gleaming gold in the last rays of the setting sun, we were all taken aback by this sudden intrusion of the wild into our lives, by the wondrous spectacle of instinctive migration.

Since the dawn of history, man has observed the periodic presence and absence of numerous bird species in any given place, noting their arrival in spring and their departure with the first autumn chills. The strident calls and agitation of cranes before "crossing the impetuous ocean" is described in the *Iliad*. The most famous naturalist in antiquity, Aristotle, evoked the periodic movements of winged fowl, and the sparrow hawk's autumn departure for southern lands is described in the Bible.

But man also saw that with the onset of winter, a host of other creatures go into hibernation, sinking into a deep torpor from which they awake in spring. Frogs, lizards, and certain mammals—dormice and marmots, for instance—retire to a carefully chosen nest to sleep through the winter. So why don't birds do the same? In the Middle Ages, and even much later, it was believed birds went into hibernation in autumn. Certain chroniclers recounted the discovery of huge heaps of inert yet living swallows brought up in fishermen's nets. Despite being keen nature watchers, these writers believed the birds had immersed themselves in the marshes, and such howling errors, in utter contradiction with everything hitherto known about the physiology of birds, would endure for many a century to come.

Even the illustrious Georges Cuvier gave us to understand that swallows can spend the winter asleep at the bottom of some marsh. But this belief was forgetting one of the fundamental characteristics of all living creatures: their mobility, a faculty that many use to ensure they are constantly in conditions favorable to survival year-round.

Their movements differ greatly. The only ones we can call migrations are those taking place periodically, usually annually, between two regions frequented at different times of year: the animal's "homeland," or breeding grounds, and another where it takes refuge during the season that is unfavorable in the former.

Flock of snow geese taking flight,
Cap Tourmente, Quebec, Canada.

Many animals move with the seasons: certain butterflies across continents, many fish and whale species from sea to sea, antelopes and elephants across the savannas of Africa, and caribou over the Arctic tundra. Yet birds are by far the most universally known migratory creatures. Peerless voyagers, they travel fast, far, and wide with great economy of energy, winging their way through the air, which requires far less effort than trudging overland. Some cover only short distances, and others journey from cold or temperate zones to tropical climes.

Others still, such as terns, also poetically called "sea swallows," think nothing of migrating from pole to pole, covering more than 9,000 miles a year with their graceful, delicate flight.

And how many albatrosses have sailed around the globe in the Roaring Forties, matching the exploits of single-handed yachtsmen, before returning to some minute island in the middle of the ocean to breed once more?

Our entire planet is crisscrossed by a dense network of pathways leading from one continent to another, over the world's most redoubtable oceans. Countless birds of all species take part in this sumptuous planetary ballet, but its routes are littered with corpses. Millions fall by the wayside, victims of storms, sandstorms, exhaustion, or predators, paying the price for the survival of their species.

We know more today about the migratory calendar, the routes chosen by each species, and each of their populations in function of their needs and aptitudes. We now have a clear enough picture of where birds breed and where they spend the cold months of the year, virtually always returning to these same distant locations year after year. Yet although bird migrations have gradually ceased to be a mystery to us, our fascination remains undiminished. Many questions remain unanswered, and, as always, nature will have plenty more surprises in store for us.

So, leaving the ornithologists to their captivating task of finding those answers, let us now journey through this book, accompanying the birds on their long voyages through the great theatre of the skies. Thanks to these wonderful, ineffably poetic pictures, taken without recourse to any special effects or artifice, we can catch them in full flight, muscles straining as they wing their way toward their wintering grounds or back to their birthplace, where, when spring comes, they will court and mate again. Over thousands of miles they will elegantly struggle, through storms on the ocean, over the desolation of the desert.

These pictures are the truth and nothing but the truth. They are visual evidence obtained by methods no scientist would refute. Yet they also have a formidable artistic and emotional dimension. Leafing through this book, I can almost hear the rustle of beating wings, or the myriad sounds these voyagers make as they ready themselves for the next stage of their journey. Never before have we been able to participate so intensely in this, the most adventurous phase in a bird's life.

Meadows in Lower Normandy, France.

Bar-headed geese in
Central Asia.

Barnacle geese over the
Icelandic desert.

Snow geese landing, Cap
Tourmente, Quebec, Canada.

To make *Winged Migration,* the magnificent film that inspired this book, Jacques Perrin had to put together a first-rate, cohesive, and harmonious team of movie professionals and ornithologists. He has long been acclaimed for his approach to the animal kingdom. The casts of his films, ranging from the most prestigious primates to the humblest insects of our woods and fields, have enchanted us. When he turned his mind to the world of birds, he was well aware that it would be the most difficult of all the animal realms to penetrate. Birds move in a three-dimensional universe, they are fleeting, often unpredictable creatures, and, furthermore, they go easily where we can follow only with great difficulty.

The astute observer usually finds a bird where he expects it to be, but that bird will hardly ever do what is expected of it at any given moment. The vagaries of working out in the wilderness also have to be taken into account, as do the changing seasons, which are variable from year to year, and of course bad weather, which disrupts both bird behavior and the shooting schedule. Finally, the specially adapted techniques used to follow migrations by air, to literally fly with these wild birds, must never in any way modify their behavior. Jacques Perrin and his team, whose work I had the pleasure of following, succeeded in using the most sophisticated techniques while rigorously respecting the spontaneity of their wild cast.

Winged Migration is a hymn to life and its dynamics, one transcending any mundane recording of this most adventurous aspect of the saga of birds. Jacques plunges us into the heart of the seasonal cycle of the most mobile of all creatures. He and those who worked at his side should be aware of the gratitude of all nature lovers for their wondrous achievement. A very beautiful commentary by Jean-François Mongibeaux brilliantly accompanies these sumptuous pictures.

As we contemplate such migratory voyages, the planet almost seems to disappear beneath their myriad paths. Season after season, millions of winged travelers make these journeys, and as they follow this immutable yet transient cycle, they are the very symbol of the life and dynamism of each of its manifestations.

JEAN DORST, of the Académie des sciences, Institut de France

Whooper swans migrating.

Whooper swans in the Far East.

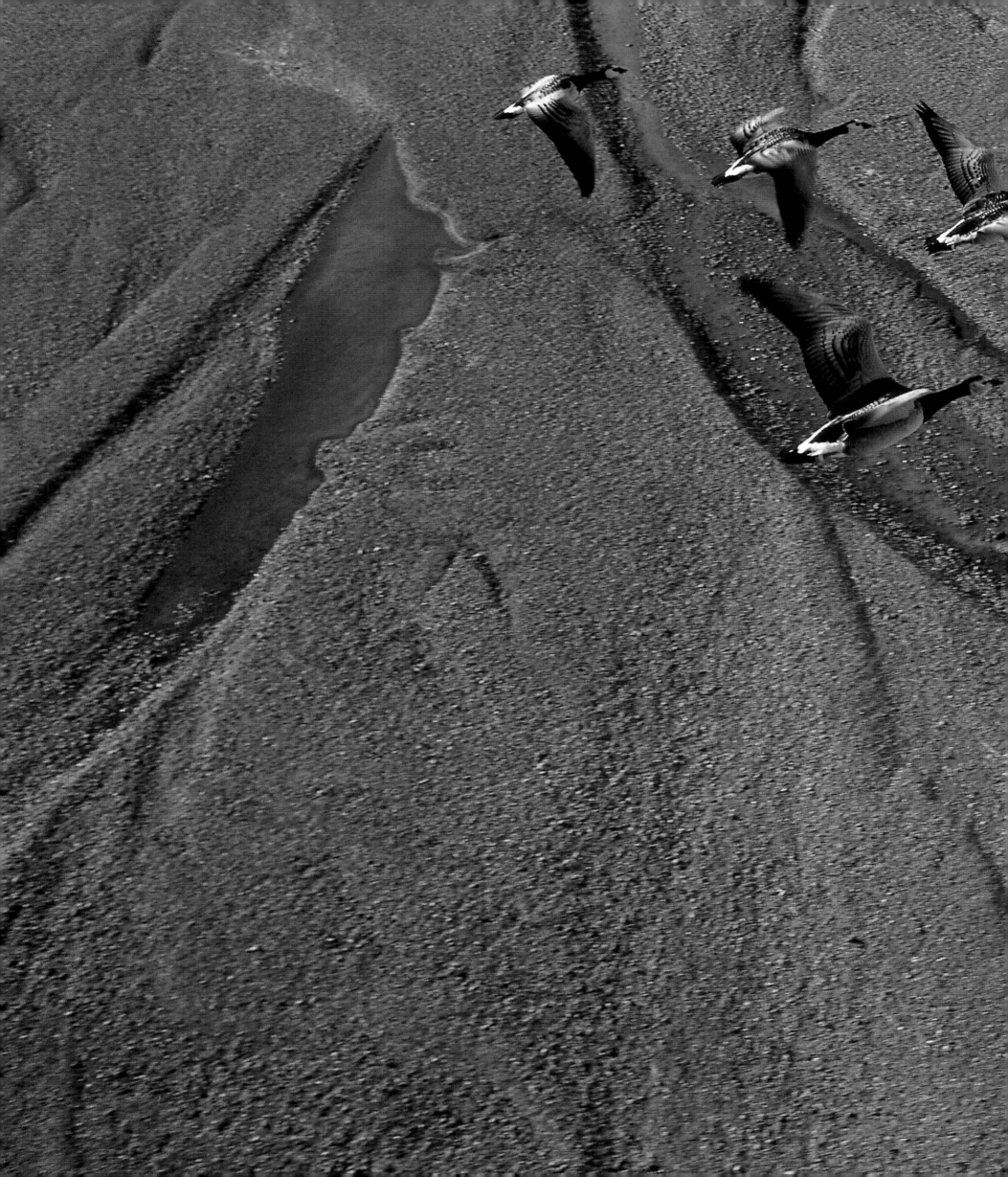

"And God created . . . every winged fowl after
his kind: and God saw that it was good."
Genesis 1:21

And then there were birds

Wandering-albatross couple
nesting, Crozet Island.

A flight of starlings on their
way to their winter dormitory
in the Jura range, France.

Gannets, Skrudur Island,
Iceland.

Gannets gliding over their
colony, Iceland.

Pink flamingos in the
Brazilian sky.

W

*ho has never
heard a robin sing
in spring?*

Crystal-clear notes announce his presence, there, on a low branch, fitfully jumping about on his perch as though about to take flight, his bright red breast swelling with each burst of song. With a few quick wing beats, he lets himself drop to the humus-covered ground. Hopping here and there on his thin legs, he pecks up a spider, then a small worm, then goes and shakes himself in a puddle, ruffling his brown wing feathers in the water until, suddenly anxious, he decides to flit back up to his perch on the branch.

Installed again in his watchtower, he returns to singing, broadcasting his presence to the four corners of the valley.

And with that song begins the story of *Winged Migration*.

Spring is here; nature is waking up. Delicate butterflies alight on the first flowers. And, for the birds—whether year-round residents or long-distance travelers like the wild geese who land in honking flocks not far from the robin's territory—the mating season is in full swing.

Already, female robins tirelessly flit here and there, bringing back tufts of grass and moss to build their nests in hedges.

Like almost every species of bird in every latitude and longitude, from the Arctic to Antarctica, in Europe, the Americas, Asia, and Africa, as soon as spring comes, they obey, each in their own way, the same law their ancestors have obeyed since the dawn of time.

And then there were birds

That robin in the hedge, our little feathered cicerone, beckons us to enter the world of birds. Proudly singing on his lookout perch, he carries out his three duties: keeping watch over his territory, a few hundred square yards of coppice and meadowland; dissuading another male robin from entering it; and, above all, persuading female robins to join him there.

Like almost every bird in the world, his song serves two purposes—to ward off males and attract females.

His call is the sound of spring, of a feathered sentinel tirelessly pouring out his song of love.

In the countryside all around, thousands of couples have already formed. Some, like the wheatears, have come a long, long way to breed here, returning to the same ancestral spot their parents did before them. The geese who landed in the meadow to rest will journey north to the nesting sites members of their species have been returning to from time immemorial. And now, snug in their nest, four blind, featherless robin chicks, beaks gaping, shriek for food. Frantically, incessantly, their parents come and go from meadow to nest with small worms and insects for their famished young. Their chicks will soon increase their weight fourfold, eightfold, from one-tenth of an ounce to more than half an ounce.

Two weeks later, as little balls of feathers, they leave the nest one by one. Watched by their parents, they will hop from branch to branch until, a few days later, emboldened, they make that first tentative jump into the blue. The sky

theirs at last—or almost—the fledgling flyers flit from bush to bush, falling, picking themselves up, dusting themselves off—and then, next spring, one of them will sing his song in this very spot to begin the beguine again.

Winged Migration takes you on a great adventure, through the air, over land and sea. After their migration to their spring breeding grounds, in the autumn the birds will return to their winter homes. Over meadow and mountain, continents and oceans, they will fly by the thousands, by the millions, the paths of their successive waves criss-

Robin on the way back
to its nest to feed its chicks.

Feathered Facts

- *The ostrich lays the largest egg of all, weighing up to almost 4.4 pounds, about 1 percent of its body weight. But the little spotted kiwi, a ratite (flightless bird) from Oceania, holds the record for the lowest egg/body-weight ratio: its egg weighs 15 ounces, about 25 percent of its weight.*

- *It is the male emu that takes care of the couple's chicks, sitting on them at night and taking them for walks during the day, while the female aggressively protects the family's surroundings from intruders.*

Feathered Facts

crossing in a planetary aerial choreography. Some, like the cuckoo, journey alone, but most migratory birds of a feather travel together.

The cuckoo is one of the most mysterious birds in the world. There she is, hidden away in the foliage, resembling a small sparrow hawk, frequently changing perch, on the lookout for one thing—a nest. But not just any nest: the nest of a cheerful little fawn-colored bird called a reed warbler.

She makes a beeline for a clump of reeds, for the nest she knows is hidden there, a nest with five pale green eggs in it. She doesn't have much time. Quickly, she gobbles up one of the eggs and lays hers, slighter bigger but the same color as the four others, in its place. Then, mission accomplished, in the nick of time, she returns to base. The female warbler has returned, and unsuspectingly snuggles down to sit on her five eggs.

Over the next few days, the cuckoo will keep watch over the nest to make sure everything is going according to plan, while at the same time continuing to lay an egg here, an egg there, in every warbler's nest she finds—as many as twenty eggs—and monitor them, too.

And then, her marathon laying spree over, one summer night she will leave her eggs with their surrogate mothers and wing her way south, just as the male cuckoo who impregnated her has already done.

Twelve days later, a chick hatches in the warbler's nest (usually two to three days before the warbler's own eggs)—a blind, featherless cuckoo chick that knows exactly what its first task is. In the coming hours, one by one, it will load the unhatched warbler eggs on its back and, bracing itself against the rim of the nest, tip them out. It's quite a feat for a newborn chick weighing less than an ounce to lift an egg almost as heavy as itself, a feat that takes three to four minutes.

If any of the host bird's chicks do hatch, they will die of lack of nourishment and their skeletal corpses will be thrown out of the nest, for the next

Black-throated bunting, Falkland Islands.

task of the insatiable baby cuckoo, once it has its blindly devoted adoptive parents all to itself, is to rapidly gain weight. The warbler's alien off-spring is so well programmed that, according to some observers, it can even perfectly imitate the heart-rending shrieks of its ex–nest fellows, now wasted away.

Every female cuckoo deposits its eggs in the nests of a particular species of bird: dunnock, wren, robin, wagtail, sedge warbler, various song-birds and passerines—several score species in all, each of whose eggs it imitates perfectly. This selective mimesis will be passed down from mother to daughter "warbler cuckoo" or "robin cuckoo," which will in turn engage in parasitism with the same host species, and never another.

Four weeks later, the young cuckoo, now weighing 4 ounces, will have far outgrown its ever-faithful 1-ounce foster parents, now worn thin from feeding their monster chick. And soon, one fine day, without any prior training, the cuckoo will leave the nest to practice flying in the vicinity. Then, after a discreet stay in the area, it will in turn make the long journey south—a journey it will make alone, traveling by night, a journey of several thousand miles that will take it over the Mediterranean and the Sahara to its win-ter hideaway, a forest in West or Central Africa. The following spring, the cuckoo will return, alone, by night, to where it was born, where its famous call will again ring out—a call all its own this time, one destined to attract a female that, like him, has made her way back here from the

Young cuckoo being fed by its adoptive mother, a reed warbler, Franche-Comté, France.

Feathered Facts

depths of Africa, and that he will abandon as soon as they have mated.

While the cuckoo is one of the most enigmatic and solitary birds in the world, barnacle geese are models of sociability. Each spring, these small, pale gray birds with light bellies, easily recognizable by their white faces, cross the North Atlantic by the thousands to nest in the far north.

Braving fair wind and foul, they reach the coast of Greenland, where patches of snow still cling to the tundra. Wings arched, undercarriages stretched out in front, a small group of them lands on stepped cliffs overlooking a river. Couples that have remained together since they first paired build their nests all over these rocky platforms, well out of reach of prowling arctic foxes. A few months later, as the days begin to draw in again, every goose, young and old, begins to prepare for the autumn migration. Banished from their Arctic Eden by the coming cold, entire families will take flight one morning for yet another, or their first, journey south.

On Skrudur Island, off the coast of Iceland, in every cleft in the rocks high above the Arctic Ocean, baby common murres with slender heads wait for food. They were conceived after the elaborate courtship displays of their parents, which, by the thousands, as though driven by a single force, whirled, soared, and dived into the sea for hours on end in a synchronized ballet of love.

The common murre is gregarious during breeding on the cliffs of Iceland. Each female lays a single egg on the bare rock, and the male and female take turns sitting on it for about a month until their half-precocial, half-altricial chick hatches. (See definitions on page 48.)

If this single egg, pear shaped to prevent it from rolling off the cliff edge, is not snapped up by a seagull or an arctic skua, the chick hatches with a thick coat of down and with its eyes open.

Brood of young barn swallows
almost ready to fly.

White stork snapping
its beak together before
beginning its display.

For twenty-odd days, it will be fed on small fish until the time comes to leave its birthplace. Prodded forward by the beaks of its parents, threatened on all sides by marauding birds of prey, the young murre is urged on its way to the edge of the precipice. There, for the first time, it looks down on the breathtaking spectacle of the Arctic Ocean crashing against the rocks far below. From overhangs all around it, hundreds of other young murres jump headlong into the void, though gulls and skuas gorge themselves on this providential rain of chicks, snapping them up in midfall.

The call of the sea is irresistible. The young murre, still only a tiny ball of down, throws itself out and downward. At the foot of the cliff, flapping its still-featherless wings, it bounces off the rocks into the breaking waves. Bobbing up and down like a cork in the foam, it shrieks for its parents, which, having identified their chick among hundreds, are already flying low overhead to reassure it.

But the ordeal has only just begun. With its father, and only its father, the young bird will now swim far out to sea to finish growing for about a month. There, it will learn how to fly underwater like a penguin. But, unlike a penguin, it will also learn how to rise up into the sky by flapping its wings against the surface of the sea. Gradually, its father will leave the young bird to its own devices. When it can fish on its own, he will abandon it for good.

For years, the young murre will roam the oceans alone. Only after four or five years, when it has reached maturity, will this lone child of the high seas infallibly find its way back to Skrudur, even if doing so means traveling thousands of miles. The time has come for it to join its fellow kind in the vertiginous aerial ballet imposed by murre courtship protocol, and, again, no challenge will have been too great for the survival of its species.

Feathered Facts

- *The chicks of the merganser, a migratory duck, are born in a hole in a tree. Two to three days after hatching, they climb out of the nest opening and jump to the ground, where their mother waits to lead them to water. If there are obstacles on the way, she carries her chicks on her back.*

The egg tooth

All chicks inside the egg have a tiny, horny appendage on the end of their beak called an egg tooth, which they use to cut—not, as is often thought, to break—their way out of the shell. The chick takes up almost all the inside of its egg except for an air pocket next to its head, so it has very little space to work in. With its egg tooth it chisels out a very precisely shaped hole, usually oval, through which it then extricates itself from the egg.

The egg tooth falls off the chick's beak soon after the young bird hatches.

Female eider by its nest,
Bylot Island, Canada.

Feathered Facts

• *The house sparrow, one of the most prolific wild birds, can rear up to five broods a year. But it is the domestic duck that lays the most eggs, several hundred per year, followed by the domestic goose with about fifty. Albatrosses, penguins, and large birds of prey, by contrast, are single-egg layers. As a general rule, the harsher the living conditions, the fewer eggs birds lay, and the birds of species that lay the least eggs are often those that live the longest.*

Common murre.

Ibis in Southeast Asia.

Killer chicks

The honeyguide, an African species, lays in other birds' nests, usually those of hole nesters such as woodpeckers. As soon as honeyguide chicks hatch, they ferociously attack their nest mates, inflicting mortal wounds with the razor-sharp, vampire-like egg tooth on their upper bill. Having killed the host bird's chicks, the baby honeyguide then receives the undivided attention of its adoptive parents, which bear not the slightest grudge, and the needle-like tooth, having served its purpose, falls off two weeks later.

Perfect timing

Short-tailed shearwaters, which nest in gigantic colonies along the coast of Australia, lay their eggs over a very short period (about twelve days) at the end of November. Every shearwater colony, parents and young alike, then musters at exactly the same moment to migrate across the Pacific. Due to this extraordinary biological synchronicity, hundreds of millions of these birds cross the largest ocean in the world at the same time.

Baby-sitter birds

We know female ducks will readily take care of the brood of an ailing mother. This behavior is widespread in birds, with parents of one species often taking care of chicks of other, sometimes very different species. Mutual aid, too, practiced by species ranging from pigeons to swallows and tits, usually involves feeding and educating young but can also include nest building. In the case of the magpie goose, an Australian bird, the male may have two mates at once, which lay their eggs in a communal nest and share brooding, feeding, and chick-raising duties. Conversely, the Galapagos buzzard is polyandrous, one female having up to half a dozen mates, all of whom take care of the young and defend their territory with no apparent hierarchy between them. But the ani, an American cousin of the cuckoo, practices the most extraordinary mutual aid of all. Up to ten couples build a communal nest out of twigs and leaves, with immature birds participating in this collective enterprise. The females then lay their eggs randomly inside and take turns incubating them.

Feathered Facts

- *The chicks of the whydah, an African bird with habits similar to those of the cuckoo, are identical to those of the host-nest chicks, usually waxbills. Even the markings on the inside of a whydah chick's beak—color stimuli that encourage parents to feed their young—duplicate those of waxbill chicks, their colors and complex patterns differing with each of the numerous species of waxbill that host them.*

Courting favor

The male of the estrildidae, a Pacific and African bird family, brings a female gifts of flowers and twigs to seduce her. During the mating season, the males of certain aquatic species—herons and grebes, for instance—bring the female nest-building materials, accompanying these offerings of branches or plants with seductive bows and feather bristling. The males of some bird families—birds of prey, terns, and tits—bring females food, and male jays, wood-peckers, and nuthatches prepare their campaign in advance, building up hoards of seeds and fruit before or during the winter to feed females during the mating season. But the most extraordinary gift bearers of all are male puffins, terns, and seagulls, which bring their mates little fish. The male does a kind of dance around the female, showing off the fish clutched in its beak and refusing to give them up until the very last moment.

Atlantic puffin
clutching sand eels in
its beak, Iceland.

*T*he breeding season, which for birds can last several weeks, begins with the staking out of a territory to host the nest and provide for the family's food needs. The more limited the resources, the more vital the demarcation of this territory, whose size depends on the abundance and distribution of food sources.

To defend this zone and demarcate its borders, male birds use various acoustic and visual signals, which they also use to attract and seduce females. These signals generally consist of calls and song but may also involve beating their wings together—pigeons, grouse, ptarmigans, nightjars, goatsuckers, and nighthawks do this. Certain species use other signals such as vibrating specially modified feathers during flight (woodcocks and hummingbirds), snapping their beak shut (a characteristic of storks), or hammering on tree trunks (a specialty of wood-peckers). And, of course, many male birds also display their plumage, showing off their feathers with specific postures and behaviors—a specialty of robins, shanks, ruffs, and birds of paradise.

The males of certain species bring females gifts such as token nest-building materials or food, a behavior that enables a female to test a male's food-seeking efficiency and at the same time provide her with the indispensable food supplements she needs to help her form her eggs.

But before building a nest and laying, male and female birds first have to meet each other. The overwhelming majority of species (90 percent) are monogamous, and large birds with long life spans, such as swans, condors, cranes, and the wandering

First, stake out your territory

by Stéphane Durand, ornithologist, assistant director of *Winged Migration*

albatross, which can live up to sixty years, mate for life. It is this fidelity—also the case with penguins, geese, gannets, and storks—that is often the key to reproductive success, since chicks of these species cannot be raised by a single adult.

In species whose males are polygamous, male birds may congregate to display together in one place, known as an arena, each within its own small court, or display area. This is particularly true of black grouse, sage grouse, great bustards, and ruffs. Rather like watching a fashion show, each female chooses the male she wants to mate with, which may often be the one most of the other females present have chosen. Polyandrous species (cassowaries, jacanas, and phalaropes, for example) are much rarer, in which case it is the female that copulates with several males, then leaves them to look after her progeny.

The nest's purpose is to protect eggs and chicks from the weather and, above all, predators. The choice of site and its orientation with respect to prevailing winds, rain, and sun, as well as the building materials used, are therefore of prime importance.

The nest can be a mere declivity in rock, sand, or gravel, such as that of the plover and the tern, or it can be an extremely elaborate piece of architecture involving sophisticated construction techniques such as stitching and weaving, like the nests of the penduline tit and African weaver bird. An impressive diversity of materials is used, ranging from mud mixed with saliva to branches and twigs, leaves, and feathers.

All birds incubate their eggs, with the sole exception of the megapodes (big feet), Australasian birds that use the heat generated by enormous heaps of

Gannet colony, Iceland.

Imperial-cormorant nest,
Falkland Islands

Snowy owl in its nest,
Bylot Island, Canada.

Southern giant petrel,
Falkland Islands.

Black oystercatcher,
Falkland Islands.

decomposing vegetation or the warmth of the sun on black volcanic sand. About eighty species, including 40 percent of cuckoo species, brood parasitically among other birds.

Certain ground-nesting species in tropical regions, such as sandgrouse and lapwings, sit on their eggs only during daytime, to protect them from the sun. The incubation period varies greatly from species to species, from about ten days among small passerines such as the tits to more than eighty days among certain albatrosses. To achieve optimum transmission of body heat to eggs and therefore improve incubation, many birds temporarily shed feathers on a precise and highly vascular area of the abdomen called the brood patch; ducks and geese pluck out the feathers covering the brood patch and line the nest with them. In 54 percent of bird families, the partners take turns to incubate, in 25 percent only the female incubates, and in 6 percent only the male does so. Among the remaining 15 percent of bird families, all three options apply.

Since incubation takes up 60 percent to 80 percent of the day, the partner not sitting on the eggs often brings food to its brooding mate. Hatching can take from twenty minutes to four days, as with albatrosses. Parents very rarely aid a chick in hatching, but always discard the shell as soon as it does so. There are two main types of chick: precocial chicks, such as ducks, geese, and all the gallinaceans, which are born covered with down and with their eyes open and are immediately able to walk, run, swim, and find food for themselves, and altricial chicks, such as robins, jays, magpies, and parrots, which are born

featherless, blind, and incapable of coordinated movement other than opening their beaks to shriek for and receive food, and which are totally dependent on their parents. Precocial birds leave the nest as soon as the last chick has hatched; synchronized hatching is therefore crucial for them, and incubation begins once the last egg has been laid. In altricial species, brooding can begin before the last egg is laid, thereby provoking successive hatchings and chicks of different ages. Between these two extremes, one observes a whole range of intermediary strategies, as in terns, whose chicks are born with down and their eyes open, but which remain in the nest. There are also very precocious chicks, such as those of the megapodes or the black-headed duck, a South American species capable of flying twenty-four hours after hatching.

Foldout:
At the end of winter, young and adult Japanese cranes dance and strut as a courting display or to reinforce the bond of the couple, Hokkaido Island, Japan.

Page 49:
Black-browed albatross sitting on its egg, Falkland Islands.

Avian architecture

As a rule, the nests of birds in temperate zones are less spectacular than those living in tropical habitats. Birds expend considerable energy building them: a chaffinch, for example, can make more than a thousand return trips carrying various nest-building materials. The barn swallow also comes and goes about a thousand times to build its nest, which is an accretion of the tiny mud balls it brings in its beak.

— African weaver birds, or malimbus, build their nests in acacia trees out of plant and wood fibers and thin lianas. Some strip lengths of fiber off palm leaves by grabbing a leaf at its base and allowing themselves to fall. Having constructed a kind of roofed hammock, they then "sew" these materials together using various knots. The monk parakeet does the same.

— The smallest nest of all is the calliope humming-bird's, at only 3/4 inch in diameter—but then its eggs weigh only an ounce.

— The white stork's nest, which it adds to year after year, can weigh up to 2,000 pounds.

— The hammerkop, an African wading bird weighing about 1 pound, builds a nest 5 feet wide and almost as high in trees. The structure, weighing around 100 pounds, is capable of supporting the weight of a human.

— The Egyptian plover, or "crocodile's toothpick," lays its eggs on the ground, then covers them with sand. The sun then heats them as it does reptile eggs. At nightfall, the bird returns, clears away the sand, and sits on the eggs during the cool hours of the night.

—The red ovenbird, or red-breasted nuthatch, builds a clay nest on the ground that has two chambers sep-arated by a partition. This two-room suite weighs almost 11 pounds, whereas the bird itself, the size of a blackbird, weighs only 3 ounces.

— The republican, a South African weaver bird, builds a huge communal nest out of straw that can reach 20 feet in diameter, with a single entrance and capable of housing up to a hundred couples.

— The Australian brush turkey builds its nest out of piled-up leaves. The nest can be up to 3 feet high and 6 or 7 feet across and can weigh 2 to 4 tons.

— In Asia, the common tailorbird uses two large tree leaves, which it folds and assembles by piercing small holes through which it threads thorns, twigs, cocoon silk, or even spider web filaments. Resembling a kind of upside-down cone and firmly attached to the tree by the stalks of the leaves, it remains green and renders the nest even more discreet.

— Certain large species of the megapode family use an artificial "incubator." Instead of making a nest, these peacocklike natives of Australia, New Guinea, and the Philippines accumulate gigantic heaps of humus, whose fermentation generates sufficient heat to incubate their eggs. Megapodes in the Bismarck Archipelago, however, leave their eggs in lukewarm volcanic ashes.

— The male hornbill, an African and Asian species, encloses the female sitting on the eggs inside their nest in a hole in a tree trunk by walling up the opening with mud. He then feeds the food he brings her through a tiny, purpose-built orifice. As soon as the chicks are big enough, the female destroys the wall from inside to free her brood.

Stork in the Calvados region, France.

Oropendola weaving its hanging nest, Peru.

— The giant coot of the high Andean plateaus makes an artificial island for itself out of a "mattress" of reeds and aquatic plants, several yards in diameter, on which it builds its nest. The structure is thick enough to support a human.

— Long-tailed tits, by contrast, make their nests out of spider webs. Very elastic and capable of containing up to twelve chicks, these nests respond to the movements of the brood and adapt to its growth.

— The kingfisher burrows a tunnel into the riverbank, up to a yard long and ending in an incubation chamber where the female lays her eggs on the bare ground. Her bower quickly fills up with nauseating detritus, though, and foul-smelling whitish liquid begins to ooze out of it, giving away the nest location. And each time the male leaves the nest in search of food, he has to dive repeatedly into the water to wash himself.

— The most extraordinary nest of all, however, is that of the ptilinorhynchus, more commonly known as the satin bowerbird or gardener bird, a beautiful blue-black bird native to Australia. The male builds its nest on a platform of grass and twigs about a yard across, on which he builds two parallel "walls" running north and south. But the most amazing thing about this construction is its decoration. The male satin bowerbird adorns it with assorted objects gathered in the vicinity—leaves, fruit, pebbles, strips of plastic, pieces of glass, feathers—all of which have one thing in common: they are all brightly colored and, with some birds, even exclusively blue or purple. The bird regularly renews this decoration to maintain its vividness,

having no qualms about pillaging neighboring nests to do so. But he doesn't stop there: he colors certain parts of the nest with the juice of wild berries he crushes with his beak, and sometimes even his plumage as well, using a root or a small vegetable sponge that he weaves and then uses as a brush.

So, why go to all this trouble? Because the nest's sole purpose is to attract females in large numbers, each of whom inspects several of these love nests, makes her choice, copulates inside with the nest owner, and flies off to lay elsewhere in the forest.

Nearly all birds that nest in cavities or enclosed nests generally lay white, quasi-spherical eggs, whereas those that nest on the ground or on rock have dark-colored, camouflaged eggs that are often pear shaped to prevent them from rolling, or, in the case of species such as the guillemot, falling off the cliffs where they nest.

Atlantic puffins
on a cliff edge, Iceland.

Avian lovemaking is an acrobatic affair. To achieve his ends, the male bird has to do a balancing act on the female's back while both of them flap their wings to maintain their precarious union for a few seconds, the female raising her tail, the male lowering his. It's brief, awkward, often interrupted, and may have to be repeated as many as a couple dozen times before the male accomplishes his mission. Birds usually copulate on the edge of the nest or very close by, a notable exception being swifts, who mate in midair.

Birds copulate by briefly placing their cloacae—the orifice of both their excretory and reproductive systems—in contact with their mate. Except for a few species such as geese, ducks, swans, the southern cassowary, kiwis, and emus, males have no penis. The ostrich holds the record with a phallic appendage that can be up to 8 inches long, and is retractable. The minute red-billed buffalo weaver of Namibia was recently discovered to have a small appendage originally thought to be a kind of penis—but it has turned out instead to be a lure affording the male quasi-continuous excitement; the females of this species also have this appendage, though it's not quite as long as the male's. Researchers believe the purpose of these strange organs is to produce an erotic friction during copulation.

Do birds in fact experience sexual desire or pleasure? In males, at any rate, the volume of the testes can increase five hundredfold, and the weight of the testes four hundredfold, during the breeding season. The males of some species resort to violence to get their way. For example, a group of mallard drakes may tail a duck and hold her head underwater while they

brutally rape her. An aroused male sage grouse, finding no partner during parading, can relieve himself on a clod of dirt, and the hummingbird does likewise on leaves.

Male homosexuality is extremely rare in birds, with the exception of the hooded warbler, which can adopt an effeminate pose to seduce other males. But lesbianism is frequent, particularly in silver and California gulls, Canada geese, and Caspian terns, and in western-gull colonies, about one couple in ten consists of two females, which enlist an already paired male solely for impregnation. Female couples of western gulls therefore incubate twice as many eggs as a mixed–sex couple, though curiously, according to American research, their eggs are smaller and have a lower hatching rate.

Polygamy, polyandry, ménage à trois, divorce, even transsexualism—you name it, birds do it.

The wren is an interesting case. We all know this busy little bird, forever flitting in and out of hedges, a tiny brown ball of feathers with a short, upturned tail. But the reason the male is so hyperactive is that he is a full-fledged polygamist. He builds several almost identical nests at once, which he encourages every female in the vicinity to visit, and, after they mate, he retains as many as he can house. Woodpeckers also prepare several nests in different tree trunks, but they keep only one female.

Infidelity is commonplace among bird couples. Without being genuinely polygamous, most of them, males and females, can copulate with multiple casual partners. Ornithologists have termed this practice extra-pair copulation, or EPC. We have little

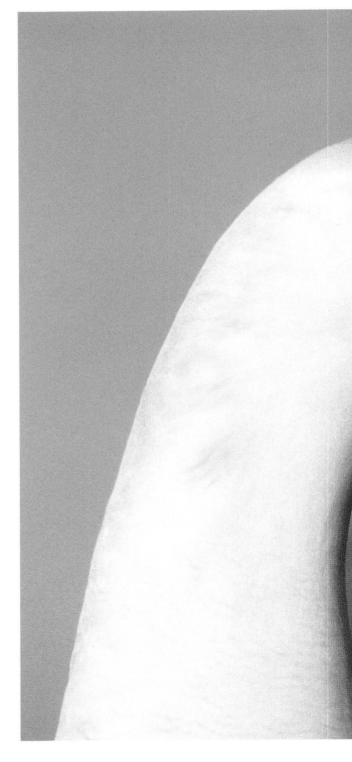

Pages 56–57:
Wandering albatrosses courting,
Crozet Island.

In some species, couples bond for life, like these wandering albatrosses on Crozet Island.

Infidelity, divorce: a bird's love life isn't always easy.

statistical information on these infidelities, but, to give one example, at least half of all mallard broods consist of offspring of multiple fathers.

Although there seems to be no jealousy among birds, some of their behavior is remarkably competitive. For instance, the dunnock, a bird often mistaken for a sparrow and, like many sparrows, an inveterate womanizer, copulates all day long with any female he comes across. Before getting down to business with a new conquest, he inspects her first. Apparently suspecting he is not the first to benefit from her favors, he titillates his partner's orifice with his beak to check whether there has been a predecessor in recent minutes. If so, which is highly probable, he evacuates the rival's semen with his beak before following suit.

Threesomes are quite rare, but they occur among oystercatchers and the Galapagos buzzard. In the case of the latter, a female can pair, often for life, with two or more males, each of which takes turns incubating eggs, keeping chicks warm, and finding food.

Is the lifelong fidelity of certain birds merely a legend? Although stork and albatross couples, among other species, do nearly always remain together for life, in many species, long-term couples can separate. "Divorces" are quite widespread among pink flamingos and barn swallows but also occur among king penguins, puffins, cormorants, larks, and most passerines (tits, flycatchers, and so on). The separation nearly always takes place on the female's initiative, and ornithologists have long believed that couples separate because of failure to reproduce, enabling them to be available to couple with other partners.

Although rare, cases of transsexualism do occur, notably in pheasants, American robins, chaffinches, ducks, partridges, herons, ostriches, and domestic hens, and transsexual mutations can engender extraordinary anatomical changes in birds. Many a farmer, for example, has seen a hen suddenly start crowing like a rooster at daybreak, and then caruncles starting to grow on its throat, and sometimes even a crest forming on its head. After a year, transsexual hens may display, and some even impregnate their ex-sisters.

Young common cranes, Aveyron, France.

Pages 60–61:
White storks fencing, Lower Normandy, France.

Incubation, or brooding?

As soon as the egg in the mother bird's abdomen is fertilized, it begins to develop, but once it is laid, the development of the embryo continues only if it is maintained at the required temperature. As a rule, parent birds maintain this temperature by heating the egg or eggs with their own body heat, an activity we call brooding. Incubation, by contrast, is the development of the embryo in the egg from its fertilization to hatching, even if the egg is not continually brooded or even not brooded at all.

Shopping for seafood

Some birds have to cover very long distances to feed their young. The seagulls of Chile and Peru, for example, which nest in the middle of arid deserts, can cover more than 50 miles to reach the Pacific. But they are far from rivaling the exploits of certain species, which can go on extraordinarily long return trips to feed their young.

King penguins, for example, which nest on Crozet Island, near the Kerguelen Islands, can swim up to 1,000 miles to reach their fishing grounds to feed their young. But wandering albatrosses hold the record. Argos satellite tracking transmitters fitted on the backs of wandering albatrosses nesting on Crozet Island have proven they cover an area from close to South Africa to Antarctica. Scientists were amazed to discover during these experiments that certain albatrosses had covered almost 1,000 miles, solely to find food for their chicks.

"How do you know but ev'ry Bird that
cuts the airy way, is an immense world of delight,
clos'd by your senses five?"

William Blake

Life
on
the wing

Red-breasted geese in
eastern Europe.

Young whooper swans on Lake
Montbel, Ariège, France.

Mallard crossing a lake,
Languedoc-Roussillon, France.

Barnacle goose.

*B*irds *may appear very
different to us mammals,
but they are among our
closest relatives.*
Like us, they live in a world of color, except for
nocturnal birds, which have black-and-white
vision. Whereas most mammals, live in a mono-
chromatic world of grays, robins, wild geese, and
albatrosses of the southern oceans see in color just
like humans do.

The eyesight of large birds of prey is infinitely more acute than human vision. High-gliding
predators—eagles, falcons, and vultures—can
make out the most minute details on the ground.
From several hundred feet up, for example, they
can tell whether a sheep lying in the valley below
is alive or dead by watching its side to see whether
it is breathing. To "lock on" to their prey and seize
it on the wing, a falcon's eye has a central "mag-
nifying glass," ten times more powerful than its
peripheral vision, enabling it to zoom in on the
details that interest it. A bird's eyesight is
superior to ours not only because it is more acute
but also because it enables a greater field of vision.
The pigeon has a 300-degree visual field, the
woodcock a 360-degree view. Without moving its
head, therefore, the woodcock can see what's
going on behind it.

Better vision requires better—and bigger—
eyes, and birds have higher eye/body-weight ratios
than most mammals. The ocular globe of a small
passerine like the starling, for example, can weigh
up to 15 percent of its body weight (the ratio
for humans is 1:100), but the ostrich beats all:
proportionally, its eyes are five times as large as
ours. In seabirds, too, the conical cells of their

Life on the wing

Greylag geese following the Seine
through Paris.

Feathered Facts

- *When severely frightened, the hazel grouse can lose part of its plumage, but most birds react to danger by blending in, and birds have perfected the art of camouflage to a consummate degree.*
- *The potoos of South America and the frogmouths of Southeast Asia and Australia, for example, conceal themselves on tree boughs whose appearance their plumage exactly imitates. The female even lays her egg balanced on a branch so that when she sits on the egg, she literally blends into her background.*
- *And when the great bittern senses danger, it straightens its long neck and points its beak skyward to imitate the reeds surrounding it.*

Mallard in mid-liftoff.

Greylag goose.

retinas, those enabling color vision, contain microscopic oily drops capable of polarizing light, and therefore of eliminating glare from the surface of the water.

Birds may well be close relatives of ours, but their hearing is far superior. They can hear sounds out of our auditory range, from bass frequencies to infrasonic and ultrasonic sounds. Nocturnal birds of prey such as owls, in particular, have extraordinarily highly developed hearing. They locate their prey in the dark chiefly by the sounds they make, resorting to eyesight only in the final phase of the kill. In addition, the barn owl enhances its already superb hearing by means of feather "trumpets" that funnel frontal sounds into its ear. It is these organic hearing aids, composed of short, rigid feathers, the product of millions of years of evolution, that give the barn owl its characteristic appearance.

If birds have such well-developed hearing, it is because they depend on it to hunt or to protect themselves from predators. But their hearing also enables them to receive vocal messages from members of their species. Birds never sing just for pleasure—they "talk" to one another. In fact, they talk to each other a great deal, as we will discover in *Winged Migration,* listening to the concerts of Canada cranes in Nebraska or wandering albatrosses on Crozet Island in the southern seas.

Contrary to what one might expect, however, birds do not use the same vocal system as mammals to express themselves. A bird's larynx plays no part in communication. Instead, it uses a

Pages 80–81:
Bald eagle about to snatch its prey
from a lake, Alaska.

Feathered Facts

specifically avian device, the syrinx—named after the Greek goddess who in legend invented the panpipe—to produce its calls and song. This strange organ is located at the point where the windpipe separates into the two conduits leading to the lungs. The syrinx, which uses the trachea like a resonating box, is very elementary in untalkative species such as birds of prey and wading birds. But it is highly developed in the songbirds par excellence, the passerines, and also in cranes and swans, which are capable of producing extremely loud sounds. In some species, the syrinx can even produce two songs simultaneously. Using the syrinx and the trachea together, birds can express themselves with an extraordinary spectrum of sounds ranging from the deep-throated roar of the bittern to the high-pitched shrieks of swallows. And these sounds carry a remarkably long way, considering the modest size of most of the birds that make them. The ringing calls of bellbirds of South America, for example, carry for more than 3 miles beneath the forest canopy.

In the 1930s, Jacques Delamain, who worked for the National Museum of Natural History in Paris, wrote a poetic (as opposed to scientific) book called *Why Do Birds Sing?* It is a question ornithologists have since gone a long way toward answering. The ancient *scientia amabilis* (science of birds) has advanced in leaps and bounds in only a few decades.

For example, we now know that, as a rule, it is the male that sings, except in a very few species in which the female also sings. We also know,

Andean condor, Argentina.

Bald eagle on the lookout, Alaska.

Pages 84–85:
The annual bald-eagle reunion on the Chilkat River, Alaska.

Feathered Facts

- *The peregrine falcon, whose cruising speed is 80 mph, can reach speeds of up to 180 mph when diving, making this small bird of prey the fastest vertebrate in the world. The common swift can reach top speeds of up to 125 mph.*
- *Pelicans are capable of flying in unison, perfectly synchronizing their wing beats.*
- *The ostrich, the fastest running bird, sprints up to 45 mph.*
- *The plumage of the hooded pitohuis and blue-capped ifrita of Papua New Guinea, the only known venomous birds, secretes a type of poison, homobatrachotoxin, otherwise produced only by South American tree frogs. Thus, these birds protect their eggs from predators by rubbing their belly on the shells.*

To eat like a bird

Birds spend a good deal of the day eating. It has been calculated that species weighing between 0.4 and 3.6 ounces consume 10 percent to 30 percent of their weight per day. A human being would have to eat almost 450 pounds of potatoes a day to equal the food/body-weight ratio of the smallest of all birds, the hummingbird.

Seabirds in particular have gargantuan appetites. The enormous colonies of gulls and terns fishing off the Peruvian coast, for example, "process" almost 5 million tons of fish in a year, and a colony of cormorants can consume 1,000 tons per day.

So, why say of people with small appetites that they "eat like a bird"?

above all, that none of a bird's vocalizations, whether calls or songs, are gratuitous. All of them have a precise meaning, and sometimes even several at once. Domestic hens are capable of making some twenty different calls, each of whose specific meanings is known to farmers. The messages of some birds, too, are understood by other species—the robin's alert call when it spots a sparrow hawk, for example. When a bird calls or sings, it might be vocally "announcing" the species to which it belongs (and each individual bird has its own vocal "signature" recognized by its entourage), indicating a food source to other birds of the same species (and also those of others), or marking out its territory and inviting females to join it there, as we have seen with the robin.

Birds speak to each other a lot, not only in their own language but also sometimes in local "dialect." According to some ornithologists, Breton seagulls, despite being closely related to Mediterranean gulls, would have difficulty understanding their counterparts. And Danish yellowhammers cannot understand German yellowhammers, researchers believe, whereas yellowhammers on the border between the two countries understand both other species.

Be that as it may, birds communicate with one another constantly, sometimes even in the dark. On October nights, we have all heard the overhead conversation of a migrating flock of blackbirds, larks, thrushes, geese, ducks, sandpipers, curlews, cranes, or robins. Migrating thrushes, for example, use cohesion calls to stay together in the dark. Anxiety calls, alarm calls, contact calls between parents and young, territorial or love songs—like the barnacle geese chattering on their way over Mont-Saint-Michel, all the birds of *Winged Migration* have a lot to say.

What about birds' sense of smell? Contrary to popular belief, they have practically none. Yet, for a long time, biologists believed just the opposite. To explain how migratory birds navigate by night, they posited that the birds' olfactory perception enabled them to smell and therefore recognize the regions over which they were flying. But experiments have never yielded the slightest proof of this.

Birds, then, have no sense of smell, or, at best, an atrophied one. Yet, here again, as always with birds, every rule has its exceptions. Certain seabirds such as puffins and petrels, and birds of prey such as the South American urubu, or king vulture, seem to have relatively well-developed olfactory systems. When the urubu flies far out at sea or over the forest canopy, this sense of smell enables it to locate carrion concealed from sight. Experiments have established, too, that the kiwi, a strange apterous bird native to New Zealand, also has a sense of smell.

And the albatross? It was long believed—and some people still do believe—that this bird has the keenest "nose" of all. One thing is sure: albatrosses are capable of locating dead fish or squid floating on the ocean surface from a considerable distance; many admiring seafarers have

Japanese-crane couple displaying,
Hokkaido Island, Japan.

Pages 88–89:
String of barnacle geese over
Mont-Saint-Michel, France.

Solidarity and symbiosis

Birds of the same species often help one another—the fishing "fleets" of pelicans is a fine example—but birds of different species also cooperate with one another. In tropical forests, heterogeneous flocks of birds often roam from place to place together, a curious phenomenon ornithologists call a "bird round." Some believe this behavior could be cooperation in seeking food at different levels of the forest canopy, the seed eaters facilitating the work of the insect eaters by disturbing their prey on leaves and branches.

Birds can also team up with large mammals; the cattle egret in Africa, for example, follows buffalo around to feed on the insects the grazing mammals disturb in the grass. As though repaying their horned benefactors, cattle egrets warn them of possible danger by special calls the buffalo can interpret, and African oxpeckers do the same with other large mammals.

Another African bird, the honeyguide, similar to the woodpecker, teams up with a small carnivore, the ratel, or honey badger. Both, as their names suggest, are extremely partial to honey, but the bird is not strong enough to open up the cavities in which bees nest, so it summons a nearby ratel by flying close to it and making special calls. The bird then guides the ratel to a nest by jumping from branch to branch, and the mammal then breaks into the nest and feasts on honey, deliberately leaving the honeyguide the wax it is specially fond of.

observed this extraordinary gift. And experiments with fish oil, which attracts albatrosses from far and wide, seem to have confirmed this. But more skeptical minds believe the albatross has no sense of smell and is simply attracted by the noise of puffins and petrels already guided to the spot by *their* sense of smell.

If birds, with only a few notable exceptions, can not smell, then can they at least taste? Birds are incontestably more partial to certain foods than others: pigeons to salt, birds of prey to giblets, pelicans to tilapias (small fish), parrots to apples, and so on. But this characteristic is often true only of pet birds or birds living in captivity. In the wild, all birds eat to feed and not for pleasure. The spectacle of snow geese munching grass on the tundra is an edifying example of this lack of discernment.

Most passerines eat insects, fruit, and seeds. But whether they be insectivores, herbivores, or carnivores, birds nearly always stick to the same diet (if one excepts certain species such as the rook, which switches from insects in summer to meat in winter). Although the question of taste in birds has not been resolved, seed eaters and insect eaters do not eat just any seed or insect, and they systematically avoid toxic parts of fruit and insects with an instinctive defense system.

Be that as it may, birds do not seem to waste time tasting food before eating it, at least if the sight of hens pecking up anything they can find is anything to go by. It seems birds choose their food

first of all visually, and anatomical studies have confirmed this: humans have several thousand taste buds, while the pigeon has only about fifty and the parrot, considered one of the most "greedy" of all birds, has only a few hundred.

If birds have little sense of smell and virtually no sense of taste, then surely they must have a sense of touch, a means of perceiving heat and cold and, above all, pain. Their nerve endings, comparable to those of mammals, are concentrated in areas of skin not covered with feathers. But here, again, we are not certain. It seems birds have very little tactile perception. One would have thought that if they needed to feel something, they would do so with the beak and feet, yet these are covered with nerveless horn or scales and seem to be quasi-insensitive, although certain specialists are still questioning this assumption. However, yet again, there are a few exceptions: certain mud-feeding aquatic species such as curlews and spoonbills, for example, have tactile buds on the end of their beaks that apparently enable them to feel their way around the riverbed.

Birds have a hypertrophied brain and other highly specialized organs that enable them to perceive the world and to move through it better than any other creature. The first of these organs is a "built-in" flight stabilizer. We have invented ingenious systems, particularly in aeronautics, astronautics, and optics, that allow us to con-

Japanese-crane couple,
Hokkaido Island, Japan.

stantly adjust our position in space: trim correctors, flight stabilizers, inertial guidance systems, and so forth. But birds are equipped with an anatomical device ensuring all of these capabilities naturally. Called semicircular canals, these organs exist throughout the animal kingdom but in an often atrophied form (except in fish). Located in the occipital bone on either side of a bird's skull, these intercommunicating canals are full of a liquid substance that orients it to its surroundings during flight. This system of microscopic "communicating vases," linked to highly innervated zones, tells the bird within a millionth of a second what its three-dimensional position in space is.

Feathered Facts

- *A hummingbird's heart beats 500 times a minute, compared with man's 80 to 90 heartbeats.*
- *The whooper swan has the record number of feathers (up to 30,000), while the hummingbird has only 900.*
- *The bar-tailed godwit uses its long, tactile beak to probe into sand on beaches to locate the marine worms it feeds on.*
- *The oystercatcher very quickly inserts its beak into a shellfish's valve before the shellfish can shut, then forces the shell open again by twisting it.*
- *The bat falcon hunts bats at dusk with great skill, swallowing them whole in midflight.*

A short history
of ornithology

With only 10,000 species, birds are the least diverse group in the animal kingdom, which totals, by one estimate, some 1.5 million species. But despite their modest diversity, birds have fascinated humans since ancient times. Yet their study, once called scientia amabilis, evolved little from Aristotle to Buffon. It was long believed, for instance, that birds flew to the moon because they could not cross the oceans, and that large birds carried smaller birds on their backs on journeys. The fascinating origin of the name "birds of paradise" is a perfect illustration of this often poetic ignorance. The first stuffed specimens of this Papuan species to reach Europe in the late eighteenth century had no feet. Naturalists therefore concluded that these birds never landed on the ground and theorized that they fed on dew and remained in the air so as to not dirty their wonderful plumage—and were thus "birds of paradise." The truth, however, was more mundane: before sending them to Europe, Papuan taxidermists, considering their feet uncomely, had quite simply cut them off.

Nearly everything we know today about birds, the fruit of painstaking research conducted all over the world by ornithologists of various disciplines, dates from the late nineteenth century. In recent years, they have made spectacular advances in our knowledge of birds, particularly in the field of migration.

Birds also have a system that can detect Earth's magnetism. (This complex subject is discussed by Professor Francis Roux in chapter 3). Many specialists are researching the mysterious "gifts" birds use to orientate themselves so precisely during migration. Do they do this, at least in part, by sensing the planet's magnetic field? Fish have electroreceptors that do the same job. A dozen bird species, including homing pigeons and, more surprisingly, robins, are sensitive to magnetism; experiments in magnetically insulated buildings have proven this. But although researchers have not yet located these electroreceptors in birds and therefore lain the question to rest, they have discovered tissues that are sensitive to magnetism and, in the brain particularly, magnetite particles and other molecules serving the same purpose. But some specialists consider that the answer may also lie in their retinas.

And last, but not least, of the anatomical singularities that make birds the most gifted creatures in the animal kingdom are the extraordinary Herbst corpuscles, sensors all over the body that render them hypersensitive to atmospheric pressure and vibrations in the air, the ground, and water to the extent that they can feel explosions hundreds of miles away.

Although they are wonderfully equipped for seeing, "talking," and hearing, but apparently with little or no sense of smell or touch, birds do not seem equipped for earthly pleasures, at least in our sense of the word. Even during their brief bouts of lovemaking, usually more conflictual than amorous, as demonstrated by the quasi-torture inflicted by mallard drakes on their mates, they hardly give the impression of being in seventh heaven.

And, yet, birds seem ever happy to be alive and keen to seduce and reproduce. The most extraordinary examples of this lust for life are their courtship displays. Even the smallest or least brightly colored birds indulge in extravagant courting behavior: the robin pumps up his chest, the sparrow shows off his dark throat, the blackcap ruffles the feathers on its head. Some birds take these displays to paroxysm.

Winged Migration gives us numerous examples of these extraordinary winged ballets, depicting species as diverse as western grebes, Sarus cranes, Japanese cranes, whooper swans, and black grouse. The latter, a magnificent bird in the gallinacean family that is present year-round in the Alps, is a good example of sexual dimorphism, the differences in appearance between males and females. The male has black plumage with blue reflections, voluminous red "eyebrows," and a white lyre-shaped tail. His sheer splendor eclipses the female, who is smaller and reddish-brown, with no serrated tail.

Male grouse, being polygamists and seemingly having a high opinion of themselves, seem to live for their courtship displays. As soon as spring comes, they take up their traditional positions in small clearings with precisely fixed boundaries. Some birds even clear this space of

Whooper swans near the
Chinese border.

Feathered Facts

- *The females of certain species have more colorful plumage than the males. Female painted snipes (rostratulidae) of Africa and Southeast Asia and the arctic phalaropes, for example, have much more shimmering plumage than the males (whose job it is to sit on the eggs and rear the chicks). Sometimes the plumage of both sexes can be equally beautiful, though. The male Australian eclectus parrot, for example, is a superb green and the female a dazzling red (naturalists once thought they were separate species).*

- *Starlings, like many other species, take "ant baths" (a behavior called "anting"), literally rolling in the insects. Some ornithologists theorize that they do this to gather provisions, which they then eat elsewhere. Others believe they engage in anting to cover themselves with tiny "cleaners" to rid themselves of parasites, or that the ants' formic acid acts as an insecticide.*

- *Procellariforms project a sticky, foul-smelling liquid, secreted by their stomach, onto their aggressors.*

twigs and small branches, pushing them away with their beaks to prepare the theater for their extraordinary courtship presentations, sometimes with scores of birds assembling to display at once (up to several hundred sage grouse!).

Each black-grouse cock fans its tail and spreads its wings to reveal his snow-white undersides. Hopping up and down, cheeks inflated, he sweeps the ground with his flight feathers, then starts turning on the spot, loudly yelping and whistling, his fanned tail folded back over his head. Very excited ones jump in the air, beating their wings and shaking their head. Hen grouse soon gather around the edge of the arena, cackling to encourage their champions. In the midst of all this cacophony, two cocks may sometimes joust with each other. A deliberate choice of adversary? A chance skirmish? A response to encroachment by one cock on another's dance territory? Ornithologists believe this behavior to be an attempt by one cock to join the strongest, experienced, eldest, and, yes, most desirable cocks swaggering in the center of the arena. Whatever the reason may be, irascible males join battle, pecking with their beaks and dealing out blows with their wings until one of them backs off.

A hen sometimes enters the arena, to be fussed over by every male she encounters, each of whom bows in front of her, lowering his tail. Some even prostrate themselves completely flat on the ground. The hen, having chosen one of the cocks by crouching in front of him, then allows herself to be mounted.

The paraders most sought after by hens are the ones in the middle of the arena—that is, the eldest, some of whom are given the opportunity to copulate up to sixty times an hour! In the sage grouse, a North American species, one male may be solicited for three-quarters of all copulations in a given arena and as many as half a dozen in only a few minutes while other, less fortunate males wait for him to pause for breath, or take advantage of a moment of inattention, to surreptitiously benefit from a female's favors.

His job done—and, as always, done quickly—the male loses all interest in the hens he has impregnated. Not in the least bit offended, the females return to the nests they have built nearby to lay their eggs (between six and ten), and then, on their own, raise their chicks, who will never know their dandy of a father.

The displays of the black grouse would top most bills, but the circus of birds includes many more dazzling courtship acts. For instance, there are the dances of mallards and Sarus cranes; the jousts of the ruff, which fan out their long neck feathers as a shield; the fantastic aquatic dances of the western grebes, which "run" on the water in

Greylag geese.

couples like figure skaters; the vertiginous aerial duels of eagles, which lock talons in mid-flight; and the ecstatic concerts of wandering albatrosses, which "kiss" by snapping their beaks together. Why such spectacular exhibitions? If you were to succumb to anthropomorphism, you might imagine them as a kind of beauty contest in which the most attractive males win females by their greater splendor. But ornithologists claim this is nothing of the sort. If a female sets her heart on the male that has the most shimmering plumage and the most impressive dance, it is simply because these characteristics prove his vitality and health. And, if the same male is chosen more often than others, it is to better ensure the species' survival.

Two stories about bird seduction:

The first anecdote involves an ornithologist who noticed that female swallows prefer males with a long tail. To set his mind at rest about this behavior, with a pair of scissors he snipped off a tiny bit of the tail feathers of a male he knew to be greatly in demand with females. And he was right: they ignored him. He then stuck the missing ends of the caudal feathers back on, and females immediately began calling on the male's services again.

The second account concerns a North American species called the red-shafted flicker, whose males and females are almost identical, their only difference being the "mustache" the males have on their beak—red or black, depending

Snow geese landing.

Foldout:
The great biannual mustering of hundreds of thousands of snow geese on Cap Tourmente, Quebec, Canada.

on the subspecies. As this attribute seems to attract females, a researcher had the idea of painting a false mustache on the beak of a female redshafted flicker. Its male, no longer recognizing its mate, promptly banished it from their nest. The ornithologist then caught the female again and erased the mark, and everything returned to normal.

Is it instinct or example, nature or nurture, that dictates such seemingly bizarre behavior? By what mystery of nature can a warbler chick, raised in a soundproof room, separated from other birds of its species, sing just as harmoniously as his fellow warblers in the wild, which he has never heard? And why, conversely, does the chaffinch have to be taught its species' song by its parents, being incapable of singing it if he has been separated from them? Why, on their first migration, are some young birds capable of navigating without their parents' help, while others, geese and swans in particular, are for the most part incapable of migrating alone without being previously tutored by birds of their species?

Instinct, or apprenticeship? Perhaps both, —but, if so, in what proportion? As with all vertebrates, with birds, acquiring behaviors is in fact a mixture, but far from a uniform one: inborn behavior in one species may be acquired in another, and vice versa, as we have seen with singing.

Professor Jean Dorst sums up the dominant opinion as follows: "Apprenticeship certainly plays a role in navigation mechanisms, but it is far from

being primordial. Inborn factors are manifestly determinant." This view is shared among prominent scientists. Peter Berthold at the Max Planck Institut at Radolfzell, Germany, for example, believes that for migrating birds, the decision to leave, the length of the flight, and its direction are all inborn. All the same, the greatest mysteries remain those concerning migration, as we will now see on the trail of the cast of *Winged Migration*.

Color coding

The sunbittern, a Central American and Amazonian bird, has "eyes" on its wings, which it displays to impress adversaries. In the nest, too, color plays a crucial role. The inside of most chicks' beaks are highly colored, and their individual markings, sometimes consisting of geometrical patterns, are like identity tags for their mothers. Seagull chicks signal for their parents to regurgitate food by pecking the red mark beneath their beak, just as if they were pressing a button.

Mallard crossing a lake,
Languedoc-Roussillon, France.

"Bird-brained?"

The question of bird intelligence is one ornithologists treat very cautiously. A considerable body of research exists on the subject, notably on brain size, and a great many publications have made sensational claims.

"What a chimpanzee can do, a bird can do just as well, and often better: its technical skill and dexterity are often far greater, and its vocal language is very complex," maintain Bernadette and Rémy Chauvin, biologists specializing in animal behavior and advocates of the Koehler school, named after the German scientist who proved birds' cerebral capabilities in the 1940s. For the Chauvins, the parrot and crow families are the most intelligent, followed, because of their navigation faculties, by migratory birds.

As for the parrot's intelligence, research by American ethologist Irene Pepperberg, who succeeded in "dialoguing" with a parrot called Alex, seems to confirm the findings of the Koehler school. But her results are disputed, and whether parrots are truly intelligent remains up for debate in ornithological circles.

The behavior of some birds definitely does raise questions, though. Seagulls, voracious eaters of other species' eggs but often incapable of breaking them, fly off with them in their bills to drop them on rocks. Some birds use stones to break shells, as do crows to crack walnuts and hazelnuts, and the Egyptian vulture takes a pebble in its beak to break open extremely hard ostrich eggs. The great spotted woodpecker bores into nuts by wedging them in an improvised vice in a crevice or between stones. But it is the gypaetus, a bird of prey related to the vulture, that is the most extraordinary. This bird drops bones and partially dismembered carcasses on rocks so it can extract the marrow, doing this as many times as it takes to succeed.

Birds can also be surprisingly dextrous. They are capable of building extremely sophisticated nests, and excavating complex tunnels in the ground and in trees, and of tying knots and weaving. Some, like their rock-wielding counterparts, even use plant matter as tools. To extract juice from flowers, for example, some exotic birds, such as the Darwin's finch, a native of the Galapagos Islands, use a kind of pipette. Not having a long enough tongue to extract the larvae it feeds on, the Darwin's finch finds a twig or a thorn of the right size to do the job. If it cannot find one, it breaks one off a tree and may even carry this instrument on to the next food location.

The behavior of the mallee fowl, a megapode nesting in arid regions of southern Australia, is also puzzling. For almost a month, it digs a hole in the ground about 10 feet wide, lines the bottom with leaves and plant debris, then covers this compost with a layer of dirt or sand. The female then enters via a purpose-built access tunnel and lays her eggs inside this mattress of sorts, whose fermentation will generate the heat for incubation. This is when the mallee fowl's fascinating behavior manifests itself. Rain, wind, or shine, the bird ensures that the eggs' temperature remains constantly at 91° Fahrenheit. To do so, it not only has to be capable of measuring the temperature, which it does using tactile papilla on its beak like a kind of thermometer, but also of taking into account the difference in heat between day and night and even the increased humidity after rain. If it considers the temperature in its organic incubator to be too high, it uncovers the eggs, or it adds more compost material when the temperature is too low.

One could cite numerous examples of equally extraordinary bird behavior. Remember that many species hoard provisions for a rainy day. Shrikes, for instance, impale their prey—insects, small reptiles, and rodents—on a thorn, or even on barbed wire, to create a larder. Many bird species gather food for the winter, hoarding it in caches they are capable of finding months later, even beneath thick snow.

Intelligence? Instinct? Apprenticeship? Ornithologists prefer to call these behaviors "cognitive capacities."

Blue and gold macaws
in the Peruvian sky.

"The ultimate flying machine"

by Guillaume Poyet, ornithological adviser for *Winged Migration*

All birds are of a feather in one respect: they have feathers, those supple, light, strong appendages so perfectly adapted to aerial locomotion. Known as pennae, these feathers form the lift surface that enables flight. They consist of a central quill, or shaft (calamus), from whose tapered part (rachis) sprout a series of paired branches (barbs), which in turn have branches (barbules) that interlock with those of adjacent barbs by means of tiny hooks. Together they create the dense, very thin, and extremely smooth fabric that forms the curved surface (vane) during flight.

Birds have undergone all kinds of anatomical transformations designed to reduce body weight while retaining strength. The skeleton comprises only about 10 percent of the weight of a bird (only 4.5 percent in pigeons). The weight of flying birds varies enormously from species to species: from as little as less than 1 ounce in hummingbirds to more than 30 pounds among swans and condors.

The evolution of birds' forelimbs into wings entailed the disappearance of certain bones and the amalgamation of others. Unlike mammals' bones, which contain marrow, most of a bird's bones, including the skull, are hollow (pneumatized) to reduce skeleton weight to a minimum. They owe their skeletal strength to the combination of certain dorsal vertebrae, the collarbones, the wishbone, and bones in the lumbar region. The sternum is very large and is reinforced to withstand the considerable tensions produced by the muscles attached to it, those responsible for wing beating. For optimum aerodynamic equilibrium, the main internal organs are grouped near the bird's center of gravity, above the sternum. Birds also have a specially adapted respiratory system consisting of air sacs that ensure the efficient breathing necessary for the high metabolism and rapid oxygenation of muscles during flight.

Due to the shape of the wing, convex on its upper side and concave on its underside, air moves faster over the wing, creating pressure underneath, which produces the natural sustentation that enables a bird to remain airborne. Forward propulsion is produced by wing beating.

A bird's wing form is determined by its lifestyle. The three main types of flight, more or less dominant depending on the species, are wing beating, gliding, and soaring. Wing-beating flight, the most energy intensive, is the most widespread. Ducks, geese, swans, blackbirds, thrushes, and passerines all beat their wings to propel themselves through the air. Various species use different kinds of wing beating, depending on their wing form and weight: continuous (ducks), interspersed with periods of gliding (crows), or undulating, when the bird descends with its wings folded to annul the drag between two series of wing beats, each of which elevates it anew (woodpeckers).

On long journeys, certain species such as geese and cranes form groups to fly in V formation, a configuration enabling each bird to "surf" on the

Young snow goose.

Pages 102–103: Whooper swan crossing a lake on the Chinese border.

ascending eddies produced by its neighbor's wing beats and therefore fly with less effort. Some birds, such as the cormorant, fly in single file for the same reason.

Wing-beating flight can be extremely stationary: The bird with the fastest wing beat is the hummingbird, which can beat its wings up to eighty times a second when hovering in front of a flower to feed on its nectar. Mallards on the other hand can beat their wings only five times a second. Other birds, such as kestrels, kingfishers, and terns, hover to locate their prey before diving on it. Gliding is a passive form of flight during which allows a bird itself to be carried by the air without beating its wings. The larger a bird's wing surface, the greater the distance it can cover without beating them. The pigeon, usually a practitioner of wing-beating flight, can descend 300 feet and glide more than 30 feet without beating its wings, while true gliders such as birds of prey can easily glide more than 300 feet, and the albatross can glide twice that distance.

Soaring is merely gliding upward on thermals or ascending air currents, which can reduce the energy spent remaining airborne by more than 95 percent compared to wing-beating flight. This is one reason many gliding birds of prey do not feed during migration. Swainson's buzzards, for instance, fly from Canada to Argentina without eating.

The two principal types of glider—land gliders and sea gliders—use techniques adapted to exploit the specific air currents produced by their habitat.

Over land, thermals are created by the sun beating the ground and raising air temperature. especially over open spaces. Hilly and mountainous terrain, where wind is forced upward. also create invisible thermals. These hot-air elevators enable large birds of prey and storks and pelicans to rise and travel using minimum energy. During migration, gliding birds, which usually have long, wide wings, can fly all day long, hopping from one thermal to another. The absence of hot air currents over the sea forces certain large gliders to remain over landmasses and keep water crossings to a minimum. This is why, during the migration season, we are accustomed to seeing huge flocks of hundreds, even thousands, of gliding birds "hopping" from one continent to another at strategic points such as the Straits of Gibraltar, the Bosporus, and the Isthmus of Panama, where oversea flight is minimal.

Marine gliders such as albatrosses, puffins, and frigate birds use their long, thin wings to ride ascendant air currents created by the wind, and the albatross exploits the increase in wind power from the sea's surface upward. At an altitude of around 30 feet, it glides downwind, losing altitude until, skimming over the waves, it veers back into the wind to soar upward again into the stronger wind higher up. Merely by changing direction, but always with its wings outstretched, albatrosses can cover hundreds of miles without so much as a wing beat. In order to fly like this for hours without wasting energy, keeping their wings spread, these

long-distance glider-soarers have small bones on their back that "lock" their wings in an outstretched position.

Most birds fly at between 25 and 75 mph. During migration, small birds can reach top speeds of 30 to 45 mph and large birds 35 to 75 mph.

Flight is of course an integral part of various bird behaviors: courtship displays (the synchronized and parallel flight of the male and female light-mantled albatross); amorous pursuits and talon locking in numerous birds of prey; territory defense and food seeking (more than 5 percent of bird species, such as swallows and flycatchers, procure their food in flight); the silent flight of nocturnal birds of prey, rendered possible by the downy upper side of their wings; escaping (the sudden "flight" of the pheasant) and cohesion (starlings and limicolous species often fly in closely knit, synchronized flocks that move almost as one bird, a primary function of which is to protect them from predators such as falcons).

White pelican.

Pages 106–107:
Black-browed albatross in the Howling Fifties.

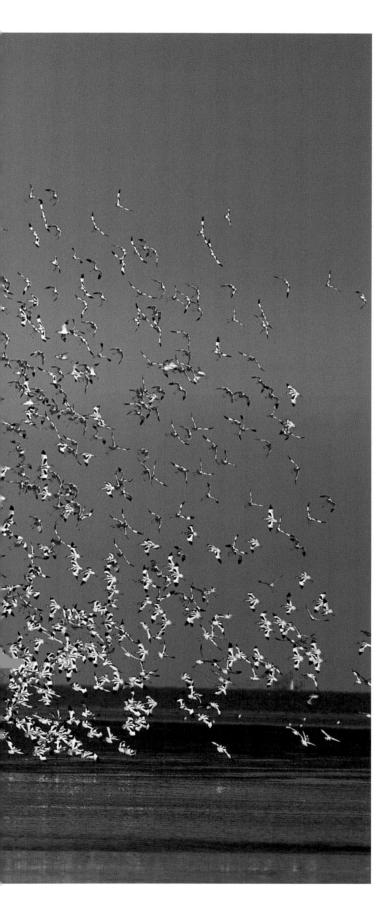

To sing like a bird

by Philippe Barbeau*, chief sound recordist for *Winged Migration*

Late one afternoon in early February, installed in my blind deep in the heart of a forest in Extremadura in western Spain, I made the final fine adjustments to my tape recorder, which was linked to two microphones concealed beneath trees a good hundred yards away.

To record the sound for a sequence of Winged Migration, *I waited for the noisy arrival of hundreds of thousands of very wary woodpigeons. Every winter, these migratory birds take refuge in this remote part of the forest.*

All quiet on the sound front so far, until I suddenly heard the melodious, unmistakable song I immediately recognized as that of the mistle thrush. Even though winter was still with us, the melody announces that the shortest days are behind us now and spring is near.

And, the moment I hear it, the mistle thrush's song takes me back to my youth in Normandy. As a teenager, in love with nature, I used to roam the forests listening for the slightest sound betraying the presence of a bird my novice eyes hadn't yet spotted. Then, as now, those short phrases of fluty notes enchanted me. The song of the mistle thrush, ringing through the early morning silence of those frosty woods, symbolized all the promise of discoveries of wild animals and vast natural spaces to come.

Associating memories with birdcalls is something many of us do in very different ways. We begin listening to birdsong in infancy, and, often without our realizing it, birds are a constant part of our everyday life, even in city centers. Their songs and calls are part of humanity's collective subconscious. The shrill calls of swifts whirling in our city skies conjure memories of hot June and July evenings. Even

when heard far inland, the shrieks of a flock of herring gulls evoke visions of fishing ports and seaside holidays. And there are so many more examples.

For the curious novice, bird-watching is often the first step in broadening one's knowledge of wildlife. And, as firsthand observation is often foiled by all manner of obstacles (dense forest vegetation, the reedy fringe of a pond, and so on), one has to learn to use one's ears, listen to the birdcalls heard all year long, and distinguish the more or less elaborate songs characteristic of the mating season. Often it is only through its vocalizations that we are aware of a bird's presence.

After a few weeks or months of listening, birds' different songs and calls become daily more familiar to us and help us better understand their lifestyles.

There is nothing more heartwarming in winter than hearing the short contact calls of tits doing their rounds in search of food, and nothing more satisfying than being able to distinguish their alarm calls warning when a predator is in the vicinity. There is the joy in spring of hearing chicks shrieking for food, and the satisfaction of recognizing the song of the males of different species conquering their territory and seducing their future partners. And what is a more wonderful spectacle than a flight of migrating cranes constantly calling to each other to stay together? Even only scant knowledge of birdcalls enriches one's pleasure in frequenting a garden, wood, forest, pond, or marsh.

I brought back from the shooting of Winged Migration *memories of a host of "acoustic landscapes" animated by birds: the profound silence of the high-altitude lakes in central Iceland, broken only by the long, melancholy calls of great northern divers; the soft*

Flock of avocets,
Wadden Sea, Germany.

whistle, like silk being torn, a condor's wings make as it glides along Andean cliffs; the dull buzzing, heard against a background sound of geysers, of hundreds of flamingos along the banks of Lake Bogoria in Kenya; the calls of snow geese flying over the mouth of the Saint Lawrence River; the deafening chatter of thousands of wading birds along the coast of Mauritania, and so on. Because of the images and sounds they produce, birds will always be charged with mystery and poetry.

*Philippe Barbeau, a chief sound recordist specializing in wildlife films, has worked on several documentaries. With Jacques Perrin, he worked on the feature movies The Monkey Folk *and* Microcosmos, *for which he was awarded the César for best sound. On* Winged Migration, *he coordinated the sound recording at numerous shooting locations all over the world.*

Snow geese, Cap Tourmente,
Quebec, Canada.

Gifted musicians, born mimics

Certain birds, especially passerines, are capable of emitting extremely diverse sounds in such a short time that the human ear is incapable of perceiving their complexity. The American chaffinch can create fifteen to seventeen different notes a second. The wood thrush can change frequency a couple hundred times a second. The grasshopper warbler produces sound similar to the stridulation of grasshoppers, consisting of some fifty double emissions a second. Some birds can make up to several hundred different sounds, grouped in phrases. The nightingale uses around ten notes to make up to twenty different calls per individual. Two hundred species, including the robin and the house sparrow, sing in "duo." The three greatest virtuosos in the bird world are the Bewick's wren, the northern mockingbird, and the winter wren.

Birds are gifted musicians but also born mimics. Legend has it that Mozart's tame starling could imitate his Concerto for piano no. 17. Legends aside, though, certain birds—magpies, jays, starlings, and warblers, for instance—really do have the gift of imitating the song of other species.

In America, a particularly gifted species called the northern mockingbird can mimic some thirty other species in less than an hour, and three-quarters of the calls and songs of the Australian lyrebird during courtship displays are pinched from other species. And what about starlings, mynah birds, and parrots, which can remember words? With training, some birds can even "converse" with one another in the most disconcerting manner. "This poses the delicate question of the intelligence of the parrot, which remains unanswered," Professor Jean Dorst considers.

Inspired by birds, composers such as Maurice Ravel and Olivier Messiaen have composed works using the avian repertoire from the nightingale's to the robin's. These artistic creations are, however, mere approximations, since musical instruments are incapable of reproducing the range of birdsong and the diversity of the sounds birds make.

Whooper swan.

Pages 114–115:
Oystercatchers and black-headed gulls,
Wadden Sea, Germany.

"Birds, lances raised to every frontier of man!
Going where the essential ebb and flow takes them,
riding its swell,
coursing the very flux of the firmament,
rounding more capes than in our dreams' imagining,
they pass . . .
and we are no longer the same."

Saint-John Perse

Barnacle geese, Iceland.

Barnacle geese, Iceland.

White pelicans flying over the
Djoudj National Bird Sanctuary,
Senegal.

Canada geese descending the
East River, New York.

D ay is breaking over the plain, nature quivering in the rising sun, a flight of migrating cranes passes high over the waking plain below, the broad, wavering V breaking up and reforming into an undulating line. Two stragglers, perfectly rectilinear silhouettes from their forward-pointing beak to their long legs stretched straight behind their bodies, beat their light gray, black-fringed wings to catch up. Squawks of celebration are heard here and there as the flock reunites before suddenly disappearing into a big, white cloud. As they have done every year from time immemorial, tens of thousands of common cranes return from North Africa and Spain to their spring nesting grounds on the great plains of northern Europe.

Migratory birds are as magical to us now as they were millennia ago, as much as they were to the Egyptian artist who painted red-breasted and greater white-fronted geese on the walls of his pharaoh's tomb to accompany the ruler on his journey to the afterlife. Centuries have come and gone, and the magic, the wonderment, remains: Where are they bound, those geese we see crossing the sky each spring and autumn? Where are they coming from? How long is their journey? How do they manage to find their way through the great blue yonder? And then there is that inevitable question all of us have asked ourselves deep down: Are these birds bearers of a message, a message as old as time itself, one we no longer know how to decipher?

The first surprise is that almost three-quarters of all birds migrate. An estimated ten billion birds, billions more than the entire human

White stork gliding
in the Libyan Desert.

Globe-trotters of the sky

race, migrate each year. And nearly 350 species of migratory birds nest in North America.

Every spring, our gardens are suddenly populated again by scores of birds with familiar songs: thrushes, warblers, chaffinches, redstarts, wagtails, swallows—a host of little singers that discreetly disappeared for the winter, some journeying incredible distances, such as the wheatear, a minute ball of feathers weighing just over 1 ounce, whose tiny wings have carried it thousands

of miles to reach its summer home. Barn swallows return to Europe in late March, where they stay all summer to raise one or two successive broods. Then, in the autumn, these birds, with a wingspan of about 1 foot, congregate on telephone wires to fly away south, across the Mediterranean and the Sahara, to winter together in West Africa. One of their winter "dormitories," in Nigeria, has a population of some ten million swallows. Redwings migrate north in the spring to nest in

Feathered Facts

• *In certain bird species such as geese, the whole family migrates together. In others, the parents leave before their young. Some birds—the ruff, for instance—migrate in successive waves: first the males, then the females, then their young.*

Greylag geese bathing
between migration flights.

Iceland, northern Europe, and Siberia. In the autumn, many of them return to milder climes in southern Europe. These are the closely knit flocks of small birds Europeans hear passing in the autumn night.

Let's now return to the large migratory birds, to the ones we know—or think we know—best, the geese, cranes, swans, and storks of this world.

Where, exactly, is Bylot Island? You look in an atlas, and there it is, to the north of Canada, above Lancaster Sound and opposite Greenland, a wild, deserted island covered with snow eight months each year. Every year, large snow geese leave by the thousands on their 1,800- to 3,000-mile journey from the East coast of the United States, between New Jersey and North Carolina, to go and nest on Bylot.

Each spring, these magnificent, immaculate white birds, very popular in Quebec and all over Canada and star members of the *Winged Migration* cast, return to the Arctic island where they were born. After a ten-day flight with a few long stopovers, they arrive on Bylot as the snow is melting. On the thawing tundra, hundreds of snow-geese couples that have paired for life prepare their nests, plucking down from their chests to line a shallow declivity in the ground. When the eggs (four or five per couple) are laid, each parent takes turns sitting on them, one feeding while the other heats up the nest, turning the eggs over from time to time.

The days continue to get longer and warmer. Summer in the Artic has arrived. Day and night now, the tundra is bathed by a dim, eerie glow. Three weeks later, in every nest, eggs begin hatching almost at once. Chicks emerge with their eyes open, chirping, covered with yellow down, and they leave the nest straightaway. The snow has melted everywhere now, but there is nothing to graze on. Their parents beckon them with loud calls: "Come on, it's time to go. Get moving!"

By the dozens, sometimes by the hundreds, depending on the colony, snow geese, big and small, leave on an incredible forced march that will take them to new pastures. It is an extraordinary sight to see these snow-white pilgrims waddling in single file for scores of miles. If they encounter a river or lake on their path, they swim across it, parents leading the way. It would be an idyllic journey were it not for the seagulls, arctic skuas, and arctic foxes there for the pickings, gorging themselves on these vulnerable feathered travelers.

Finally, after several days' hiking, the snow-white exodus arrives in a promised land of lush green inland pastures dotted with lakes, where at last the birds are out of reach of predators. But for the adults, yet another ordeal is about to begin: the annual molt. No sooner have they arrived in their chilly sanctuary gleaming in the dim Arctic light than they begin to lose the large remiges on their wings. It will take several long weeks for them to acquire their new plumage. When the days begin to get shorter, the snow geese leave for the south again, their young now

Haute Couture

Birds shed their feathers at a certain time of year, some even twice a year. These molts can be partial or total, depending on the species. When they are total, all the flight feathers (remiges) and tail feathers (rectrices) fall at the same time, rendering species that do so temporarily unable to fly and forcing them to take refuge in isolated and marshy locations, which are both rich in food and safe from predators. This is so with all species of ducks, rails, and geese, including barnacle geese, many of whom nest in Greenland.

Molts entail great energy expenditure in species that molt completely, constraining them to eat a huge amount during this period, and birds of the same species often disperse to different molting territories, the only notable exception being shelducks, which all molt at the same time on the Frisian Islands in the Netherlands.

The molt of many songbird species, such as blackbirds and chaffinches, is less dramatic and enables them to continue flying. Young woodcocks, sandpipers, and rails spread theirs out over the year: the remiges and rectrices fall at the end of their migratory journey and not before, so as not to lessen the flight capability of the birds. However, many long-distance migratory songbirds molt before leaving. It is noteworthy that in certain species, particularly ducks, during the molt males and females, even though they are dimorphic, can acquire identical "eclipse" plumage that renders them less noticeable during this very vulnerable period.

Feathered Facts

- *Red knots lacking fat (their flight "fuel") to finish their migration are capable of cannibalizing themselves in flight, eating viscera such as the liver, and even their muscles, in order to reach their breeding ground.*

Snow geese delving for sustenance on the Saint Lawrence riverbed, Quebec, Canada.

fully grown but still gray. Their autumn migration will take them over Labrador, Quebec's Ungava Peninsula, and the Saint Lawrence River, the longest stage in their journey back to the Eastern seaboard of the United States.

On Cap Tourmente, Canada is basking in an Indian summer. The trees are turning gold, and the geese feast on the green banks of the Saint Lawrence. Stopovers such as this can last several weeks, idyllic episodes in the life of milk and honey for snow geese on their way south. So effortless is this grazing that this protected species has proliferated to an extent that it has now itself become an ecological threat, and hunting the birds is authorized in spring.

We are now in the realm of the arctic tern. White birds whirl over a desolate coastline lashed by great rolling breakers, over icebergs gleaming blue in the twilight. With a noise like a thunderclap, a huge chunk of an ice cliff breaks off and falls into the crystal-clear water. Many people know arctic terns, those graceful, immaculate white "sea swallows" with their black "beret," bright red feet, and two long tail feathers. But did you know they are the greatest migratory birds on Earth? Some travel from one side of the globe to the other, halfway around the world, from the Arctic to the rim of the Antarctic—more than 12,000 miles twice a year, which is even more unbelievable considering their wingspan: only about 15 inches. On their yearly journey from one icecap to another, crossing the tropic of Cancer, the equator, and the tropic of Capricorn, the arctic tern is also

the creature that enjoys the most sunlight in the year: twenty-four hours a day for eight months and four months with longer days than nights.

The arctic terns we see soaring like butterflies above Arctic icebergs, buffeted by the wind, have arrived from a remote island off the coast of Iceland, where they reared their young this spring. Their chicks left their nest lined with shells or, at best, grass and driftwood, learned to fish and fly in only a few weeks, and then departed with their parents for happier hunting grounds. Amid gannets, petrels, puffins, and kittiwakes, the terns dive into the water, emerging again with small fish in their beaks. Relentlessly fishing, feeding, storing up fuel, these little snow-colored marathon flyers are fattening themselves up, toughening up their muscles of steel for the daunting return journey south.

In the clear Arctic morning, they leave in small groups. The flight of these champion athletes has a strangely uncertain look about it, yet they will advance inexorably down the North Atlantic, following the coasts of Europe and Africa and then the immensity of the South Atlantic to reach their destination: the often raging South Seas. The journey will take two months and when they reach the South African coast, or even perhaps the Antarctic rim, they will then make for the southern Indian Ocean, then the Pacific, before returning to the South Atlantic after doing the rounds of the Southern Seas. For a few months, these wing-loose voyagers will spend a second summer of sorts here, gorging themselves

Foldout:
Whooper swans wintering on Hokkaido Island, Japan.

Canada geese in the Adirondacks.

in waters teeming with fish before taking flight again, bound for their Arctic birthplace, yet again confirming that they are the featherweight migratory wonders of the world.

Many other birds take the same North Atlantic route as the terns and, like them, never make it. Migration is a treacherous business from which more than half of all birds never return. And those that do—that overcome exhaustion, escape predators, and brave hailstorms, sandstorms, and snowstorms, that fly over raging seas, scorching deserts, murderous hunting areas, and entire regions poisoned by pollution—achieve nothing short of a miracle. From departure to return, during the two migrations and wintering, the mortality rate can be as high as 75 percent.

* *
*

Feathered Facts

- *The Eleonora's falcon, which nests on cliffs on the Mediterranean coast from Morocco to Sardinia, feeds its chicks on small migratory birds crossing its territory. The mother times the hatching of her chicks to coincide with the passage of these birds.*

When the terns leave the far north, so do the barnacle geese. Some of these small geese, which we already met off the cliffs of Greenland, have run into bad weather off the European coast. Below their tired wings is a landscape of heaving, foam-crested mountains and a gray ship laboring through the storm, cleaving its way through wave after wave. Night is falling, and, struggling pathetically against the wind only a few feet above the wave crests, the line of geese comes alongside the enormous vessel. Skimming over the foaming water, disappearing occasionally behind a wave, they seem so fragile

Arctic tern.

compared to the huge ship forging its way powerfully through the sea beside them. They change course to fly round the leviathan, overtaking it (barnacle geese can fly at 50 mph).

The birds know the coast is not far, but will they reach it before nightfall? Two exhausted geese suddenly break away from the formation, turning back. Beating their arched wings in the wind to stabilize themselves, they land clumsily on the vessel's deck.

It is now dark. Its navigation lights on, the French navy ship continues on its course, rhythmically rearing and plunging with each mountainous wave. The two birds are resting on a hot-air vent, heads under their wings, saved, safe. Life on board the ship goes on regardless. On the rear deck, a heavy metal door creaks open. Startled by the noise, the tired birds get up, turn into the wind, stretch their necks forward, and take flight, their respite short lived. In the pale moonlight, they veer eastward, making for the flitting beam of a lighthouse. As in previous years, their companions will have stopped for the night in the broad estuary already visible on the horizon, and when the two errant birds land beside them on the marsh under the haloed moon, there will be shrieks of joy.

But not every migratory bird is as lucky as these two. Tens of millions never reach their winter homes, because of the elements they have to brave during their perilous journey, of course, but also, and above all, because of humans. For several decades now, the habitats where they traditionally rest, nest, or winter have become increasingly scarce, polluted, or threatened, despite the huge efforts made all over the world to create wildlife parks and nature reserves.

The Edens of the Arctic and the Antarctic, the vertiginous canyons of Colorado, the snow-capped peaks of the Himalayas, the virgin deserts of Africa, the crystal-clear waters around Greenland, and the emerald-green seas of Asia still host millions of wild birds. But migratory species do not live only in such enchanting locations. To pay

Feathered Facts

• *As soon as young swifts leave the nest, they fly away to Africa. By the time they return to Europe to nest in the spring, they will have spent two years without touching the ground— sleeping, drinking, feeding, and even mating in flight during their inaugural migration. A swift is thought to cover some 500,000 miles every two years.*

Long-tailed jaeger.

Pages 132—133:
Arctic terns and seagulls in a glacial lagoon, Iceland.

idyllic enough, plumes of vapor are wafting up here and there, giving away the lake's volcanic origin.

Paradise? If only. Looking a little closer, we see that most of the flamingos are having difficulty moving around. Their legs shaking, some of them labor to reach the shore. One of them collapses into the iridescent water, lays its long, thin neck on its back in a pathetic gesture of abandonment, and gradually sinks into the murky water. This is a scene not from a dream but from a nightmare—a horror movie in which heads with wide-open beaks and motionless red eyes suddenly surface from the waters of a poisoned lake.

In the gathering darkness, the scramble for the spoils begins. Familiar silhouettes busy themselves around the polluted water. The beaks, teeth, and talons of steppe eagles, baboons, marabous, wart hogs, and other predators rip apart dead or dying flamingos.

The pelicans have long since fled this carnage. It is a horrific vision not of some distant, primeval era, but of humanity's madness today.

* *
*

Feathered Facts

- *Migrating birds are often forced off course by storms. Disoriented petrels, deep-sea birds, have been found in the Jura Mountains in eastern France. Small American migratory birds such as warblers and Wilson's phalaropes, diverted by deep Atlantic weather depressions, are sometimes spotted on the French coast. Asian migrating chaffinches that have lost their way are regularly seen in Europe. Likewise, some European birds have ended up in Japan by mistake.*

a just tribute to these birds, you have to travel a bit further.

Beating their wings, then gliding, their long beaks pointed forward, their heads nestled between their shoulders, a flight of pelicans begin to descend toward one of their habitual African stopover sites. In front of them, the dreamlike vision of a huge lake sparkles in the setting sun. Hundreds of thousands of pink flamingos have already landed along its shores. And, as if this weren't

Enormous clouds of rails, majestic formations of geese and cranes, the day-and-night flights of millions of passerines, lone crossings of oceans and deserts, annual gatherings of hundreds of thousands of birds in the same places—migrations are among the most spectacular events in the animal kingdom. It is futile to try to generalize about

Couple of arctic terns near their nest concealed in the grass.

Arctic tern landing by its nest in Iceland.

Pages 136–137:
Flight of Canada geese in Monument Valley, Utah/Arizona.

Pages 138–139: Canada geese in Monument Valley, Utah/Arizona.

this phenomenon, for there are almost as many types of migration as there are species of birds: transcontinental, regional, local, leapfrog, nonstop, high altitude, hedgehopping. Sometimes entire species migrate together and sometimes it's only individual flocks; there are those that always take the same route and those that modify it each time.

Birds have been migrating since the dawn of time, obeying an ancient law, coming and going each year between their birthplace and more clement winter climes. They know no borders or nationality, their homeland being the place where they learned to fly. Migratory birds are the

symbol of liberty, a liberty that has always fascinated humans.

A scene in *Winged Migration* perfectly illustrates this respectful relationship between human and bird. Early one morning in late winter, on a mountainous plateau in France, a rooster crows. Smoke is rising from the chimney of the farmhouse. An old lady comes slowly out of the barn. For her, the day has only just begun, but the two cranes watching her from the field nearby are only pausing on their long journey. The woman approaches the birds, holding out a corncob in her wrinkled hand. But the cranes take flight, unlikely to return. Getting close to birds, winning their confidence, takes patience, but also a lot of love.

Months pass by, autumn arrives, smoke again rises from the chimney. One morning, the old lady has tears in her eyes: The cranes have returned. They wait for her outside, standing like statues. Carefully, she approaches them, again holding out a corncob. A miracle, something considered impossible, happens: timidly, the cranes come and peck the yellow seeds out of the palm of her hand.

Common cranes near a farm,
Aveyron, France

Common cranes exploring a marsh,
Aveyron, France.

European spring-arrivals

- In late January, the first robins, thrushes, tits, chaffinches, dunnocks, and thrushes begin to sing.

- The next to arrive, at the end of February, are the gargeny teals.

- Toward the end of March, the northern wheatear flies in. The cuckoo, which arrives in Europe from the south, is one of the first birds to sing in the new season. Its unmistakable call is supposed to mark the beginning of spring, but the true function of its song is to attract a female.

- The sedge warbler arrives from Africa in late March and leaves in August or September. The first warblers and the chiffchaff show up around the same time. The barn swallow also arrives at the end of March and in April and May and stays until October.

- In March or April the bank swallow appears, leaving in September.

- The sand martin arrives in late April and early May and leaves in August. The turtle-dove flies in around the same time and departs in August or September.

Pink flamingos, Lake Bogoria, Kenya.

Pages 144–145:
Flock of white pelicans, Bharatpur
National Park, India.

Feathered Facts

- *The wheatears of northern Europe have longer wings than those nesting farther south because they have to migrate a longer distance than their neighbors to the south.*
- *The robin nesting in northern Europe migrates south to the Maghreb, while southern European robins are sedentary.*
- *In France, robins are generally either sedentary or short-distance migrants, except for those in mountainous regions. In the south of England, female robins tend to band together to spend the winter in France.*

Marathon migrations

- *Between 9,000 and 12,500 miles: the arctic tern, the American golden plover, and the wandering albatross*
- *Between 6,000 and 9,000 miles: the Manx shearwater, the swallow, the bar-tailed godwit, the sandpiper, and the turnstone*
- *Between 3,000 and 6,000 miles: the willow warbler (weighing less than half an ounce), the Eleonora's falcon (capable of reaching islands in the Indian Ocean from the Mediterranean), the northern wheatear, and the steppe eagle*
- *Between 1,200 and 3,000 miles: the humming bird, the sedge warbler, the garganey, the ruff, and the snow goose*

Feathered Facts

- *When they reach sexual maturity at six or seven years, gannets return every spring to exactly the same cliff ledge where, if they can, they reoccupy exactly the same spot where they were born. The male arrives before his mate, always the same one, and they recognize each other by their calls.*

White pelicans on a beach, Senegal.

Pages 148–149:
Canada geese flying over the
Adirondacks.

An act of survival, migration requires birds to know the time at every moment, no matter where in the world they are. They must be able to navigate by the sun and the stars without any of the measuring instruments invented by humans. Flying huge distances nonstop is as natural to them as breathing.

Today, bird migrations are perfectly tracked and abundantly documented, yet they have lost none of their fascination. The awesome distances covered by migratory birds are among the most remarkable of all animal feats, but the questions they raise are far from being resolved.

Consider the birds of the western Old World. Every autumn, an estimated three to four billion migrate from Eurasia to the whole of sub-Saharan Africa. The population of the most numerous of these species, the willow warbler, a bird weighing about 3 ounces, is estimated at 900 million, that of the barn swallow 220 million, and that of the house martin 90 million.

On these journeys, entire species populations, including virtually all the passerines of the temperate zones of the Northern Hemisphere, cross great geographical barriers—seas, mountains, and deserts— to cover distances of more than 6,000 miles. For the Greenland wheatear, this feat first entails a 1,000- to 1,900-mile transatlantic flight before the first possible stopover in the British Isles or Spain. The journeys made by the birds of mainland Europe seem almost trifling in comparison. Not even the Mediterranean is an obstacle on the route south for

migratory birds. Gliding birds avoid flying across this sea by reaching Africa on thermals over the Strait of Gibraltar, the Sicilian Channel, and the Bosporus. Other types of birds cross the Mediterranean pretty much anywhere, even at its widest stretches.

The greatest endurance test of all is the Sahara. For once migrating birds begin the crossing, it is all or nothing. Landing in mid-desert would be fatal: tired birds would find neither food, water, nor shade there, and their dehydration would quicken with contact with the scorching ground. Thus the only way to surmount this obstacle is to cross it as quickly as possible, nonstop.

The shortest north-south desert crossing is almost 1,000 miles, but, as observation and the monitoring of birds with identifying ankle bands have shown, a great many birds make the northeast-southeast diagonal crossing, which can be 1,400 miles or more. A passerine's maximum cruising speed is 25 mph, and wind speed and direction greatly affect the length and flight time of its trans-Saharan journey. In calm weather, the crossing takes as little as forty hours flying nonstop, and a maximum of thirty hours in autumn, when the prevailing winds come from the north. In spring, due to headwinds, the return trip toward the northeast takes fifty to sixty hours, in which case birds can ride winds at altitudes of more than 6,000 feet. This is the route taken by the majority of migrating birds in March and April.

Radar tracking has shown that the bulk of autumn migration over the middle Sahara is done at a height of 5,000 to 6,500 feet, with waves at more than 9,000 feet as well. Passages have to be made at

Common cranes, Aubrac, France.

Migratory birds, the supreme travelers

by Francis Roux, emeritus professor at the Muséum national d'histoire naturelle (National Museum of Natural History), Paris

Young whooper swans.

even higher altitudes when the inversion of the prevailing winds takes place at over 13,000 feet. Flying at such heights has advantages, however. The air is cool—above 10,000 feet, it is subzero—which reduces dehydration. At these heights migrating birds can also avoid sandstorms, which are disastrous for them if they are caught nearer the ground. But, in doing so, they risk high-altitude weather disturbances that can be equally fatal. The "rain" of quail mentioned in the Bible, which saved the Hebrews from famine (Exodus 16) was none other than a migratory accident in the Sinai Desert.

Migration inevitably takes a very heavy toll. Only half of European passerines—at most—that leave for Africa actually return. In the western Sahara, when the scorching east wind called the harmattan is blowing, whole flocks can be carried out to sea to their deaths.

Yet despite these heavy losses, migration remains a means of survival for the bird populations that undertake it. It is a physiological need, whose purpose is to ensure birds the best living conditions all year-round. Except in tropical latitudes, where the weather is quasi-constant, there can be huge seasonal variations in habitat conditions. During winter in the Northern Hemisphere, in temperate and boreal zones, food sources either disappear or are severely reduced. For birds, migrating is a matter of escaping famine and benefiting from the better living conditions in tropical zones at the same time of year. But better living conditions for birds do no exist year-round in the tropics, either. North of the equator, living con-

ditions are optimal in September and October, at the end of the rainy season, which produces a proliferation of plants and insects. Conditions then regress as the dry season sets in: rivers retreat within their banks, marshes dry up, the plant covering degrades, and the insect population diminishes. Meanwhile, in temperate zones, spring returns, re-creating favorable breeding conditions.

The origin and biological explanation of migration lies in this geographical and seasonal alternation of optimal ambient conditions. It is the opportunist response of animals equipped with exceptional mobility to the cyclic fluctuations of their habitat. Capable of traveling great distances quickly, they go find, albeit thousands of miles away, what they lack where they were born. They then return to their birthplace to nest, certain of the security of a known territory whose resources have been tested by experience.

Long-distance migration is beyond the range of many species. It demands, in addition to highly developed flight aptitudes, other particular capacities, the most vital being extraordinary navigation faculties. True migratory birds know how to travel in a given direction but also how to navigate—that is, how to modify their flight route in function of their destination, to cross terrain of which they have no prior knowledge and which is often devoid of landmarks. This presupposes a sense of verifiable navigation: the ability to see astronomical reference points.

For a bird, the most familiar of these is the sun. When a chick hatches, it emerges into sunlight. Birds

Common cranes gliding in to land.

Pages 154–155:
Canada geese over Lake Powell,
Arizona.

know how high the sun is in the sky at midday at their birthplace, and hence the position of their birthplace, and as the sun's height varies according to the time of year, their body clock enables them to compensate for its apparent movement depending on the time of day and season. This interior sundial regulates a bird's entire physiological cycle, as it does for all living creatures.

When the sun is hidden by clouds, a bird's sensitivity to ultraviolet radiation, imperceptible to the human eye, indicates the sun's height. When the sun sets, other heavenly bodies rise in the sky. Birds' knowledge of the constellations has been amply proven by experiments in a planetarium, which have established that they have built-in maps of the night sky. Making marathon journeys involves night flying, of course, and most long-distance migratory birds prefer to travel by night, despite being diurnal the rest of the time. Nonetheless we are still a long way from understanding how birds use celestial reference points and what navigation instruments, apart from their eyes and brain, they use.

It seems increasingly probable, however, that senses other than sight are used conjointly. American research, for example, has proven that homing pigeons are sensitive to Earth's magnetic field. Birds, the studies have concluded, might measure the angle between the magnetic lines surrounding the planet and the vertical resulting from

gravity. In the Northern Hemisphere, the angle between these lines of force corresponds to magnetic north. The discovery of magnetite at the base of the brain of homing pigeons and several wild species supports this hypothesis and suggests that birds, in effect, have a kind of cerebral compass.

Whatever the system used, the navigation of long-distance migratory birds is amazingly precise. Banding, a technique consisting of fixing a numbered band around a bird's ankle, has proven this: passerines weighing less than half an ounce succeed not only in finding their birthplace each year but also in returning to their winter habitat—a particular marsh, garden, or clump of trees—despite the thousands of miles separating these precise spots. But, of course, migratory birds also use firsthand visual guides such as coastlines, river valleys, and mountain ranges, which they follow when the topographical features coincide with the general migration route.

In large species that fly in organized formation—geese, ducks, cranes—the flock is guided by older birds that have already made the journey and memorized the route. A social structure based on the durable cohesion of families ensures that a species hereditarily passes down to its young its knowledge of migration paths and stopover points.

Hence our joy at seeing enormous flocks of cranes or wild geese streaking noisily across the sky, a sight that leaves no one unmoved. These migratory birds that appear yearly at the same time, in the same places, are not only a magnificent spectacle— all the more moving because it is fleeting—but also remind us of life's constant renewal and its cycles in time and space.

Since time immemorial, for humankind the world over, birds of passage have been symbols of life and hope. They instill in us an irresistible lust for life and make us yearn to fly away with them into the blue yonder.

Canada geese, Franche-Comté, France.

Bands for research

*I*t is principally by banding birds that ornithologists are able to follow their migratory movements. During World War II, newly developed radar technology revealed birds' nighttime flight, and radar has since helped precisely determine the routes of nocturnal migrations. But it was the technique of banding, first by a Danish ornithologist at the end of the nineteenth century, that really enabled greater and, above all, more precise knowledge of migration patterns.

Banding consists of fixing a very light metal identity band round the foot of a bird or even a chick in the nest. The band carries the address of the banding center and the bird's identification number, and this information enables specialists to both identify an individual bird and record its movements. Banding also provides indications of a bird's age and life expectancy.

Up to seven million birds are banded all over the world each year. While only a minute percentage of banded birds is found, the information gathered is always precious. Thanks to banding, we know that a third of starlings caught in Berlin and then released hundreds of miles away returned to their departure point, and Laysan albatrosses, banded and then "relocated" 4,000 miles away, returned to their nest a month later. In addition, we have known for a long time now, from having banded 150,000 house martins in the United Kingdom, that they spend the winter in Africa.

In recent years, Argos beacons, transmitters that send signals via satellite, have also enabled the real time tracking of migratory birds such as pelicans, common cranes, wild swans, steppe eagles, albatrosses, short-toed eagles, and storks to further our knowledge of their migrations and winter habitats. However, banding is still the most reliable and least costly means of following a bird's movements.

If you find a banded bird, you may make a report by visiting the Web site of the USGS Patuxent Wildlife Research Center's Bird Band Labratory, www.pwrc.usgs.gov/BBL. Or, you can call 1-800-327-BAND (2263) with the band number, and how, when, and where the bird or band was found. If the bird is injured, take it to any birdwatchers' club, wildlife organization, veterinarian, or official institution capable of looking after it.

White stork with an identification band in a vineyard, Alsace, France.

Fellow travelers

Migration phenomena have existed as long as there have been animals and concern all manner of creatures, including the most minute. Zooplankton, for instance, reproduce on the ocean surface, then begin their long vertical migration downward.

Insects are no exception. Cricket migrations can be devastating scourges in Africa and Madagascar as they strip crops along the way. But one of the most extraordinary of the migratory insects is the monarch butterfly. Each year, it leaves Canada in huge clouds to winter in Southern California and Mexico, a trip of some 4,000 miles, only to die there: the return trip is made by the next generation. In Europe and Africa, two other butterfly species, the painted lady and red admiral migrate, as do the sphinx moth and death's-head hawk moth.

Many marine creatures, such as whales, turtles, and crustaceans—especially certain crabs—also migrate. Fish, in particular (including herring, white tuna, cod, and, of course, eels and salmon), cover huge distances in the sea and even swim up rivers.

Many land mammals also migrate. Not so long ago, the bison of North America crossed the vast prairies of the West to change pastures, just as caribous and gnus continue to do in America and Africa.

The human race, too, has always had its migratory "species"—desert tribes, nomads, itinerant gypsy communities, and so on— although today their lifestyles are increasingly threatened. In short, migration seems to be an enduring phenomenon that lies at the heart of all creatures.

Common crane riding a thermal.

Going the distance

- *Every year during its twenty-year life span, the Arctic tern makes the return trip from the Arctic to the Antarctic.*

- *Asian falcons also migrate annually, completing their 2,500-mile journey across the Indian Ocean in three or four days without eating.*

- *The bar-tailed godwit holds the record for nonstop flight, however, with its 7,000-mile journey across the Pacific from Alaska to New Zealand. Like other long-distance migratory birds, it deliberately reduces the size of some of its organs—liver, kidneys, gizzard—to lower its body weight for the journey, then builds up its metabolic machinery again on arrival.*

- *The tiny ruby-throated hummingbird, which weighs only a fraction of an ounce, migrates more than 2,000 miles from North America to Central America. On its journey, it crosses the Gulf of Mexico—a sojourn of several hundred miles— without landing.*

- *Another minuscule bird, the northern wheatear, flies from the New World to the Old World, where its distant ancestors originated (during the Ice Age, the two continents were joined at the Bering Strait), continuing to return to its primeval African birthplace.*

- *A turnstone, a small shorebird, is known to have covered more than 600 miles a day, flying almost nonstop for three days and nights. The banded bird made the trip from Alaska to Hawaii at an average speed of 27 mph, probably helped by tailwinds.*

- *The tiny sedge warbler, weighing just under 0.9 ounce before leaving and between 0.3 and 0.4 ounce on arrival, can journey 2,500 miles without landing. Small migratory birds can fly for seventy-five hours nonstop to cross the Mediterranean or the Sahara at an altitude of more than 3,000 feet.*

- *A Tahitian curlew is documented to have covered the 2,000-mile journey from the Aleutian Islands to Polynesia in twenty-five hours.*

- *A banded Manx shearwater crossed the Atlantic from Wales to Brazil in sixteen days, covering almost 450 miles a day at an average speed of nearly 19 mph.*

Arctic tern.

Pages 164–165:
Colony of marabous nesting
in an acacia, Kenya.

Pages 166–167:
Canada geese on a lake in
the Appalachians.

"Only birds, children, and saints are interesting."
Oscar Vladislas de Lubicz Milosz

They have
conquered
the planet

King penguins trooping toward
the sea, Falkland Islands.

Rockhopper penguins returning
from fishing, Falkland Islands.

King-penguin colony,
Crozet Island.

Migrating white storks on a
stopover in the Libyan Desert.

All color has gone from the prairie.

Winter. All is quiet. Where have all the birds gone? A melancholy call emerges from a stunted bush. It is a robin perched on a frost-coated branch, reigning alone over its frozen territory, singing at the top of its voice. A species that has come a long way to sing here today, the robin heralds the approach of another spring, its sixty millionth spring.

To discover the ancestors of birds, you must travel back into the depths of time. Once upon a time, several hundred million years ago, there was a world in which huge monsters with toothed beaks and enormous membranous wings ruled in an apocalyptic sky. These flying reptiles could have a wingspan of more than 30 feet. Gliding over the shores of the oceans that at that time covered nearly the entire planet, with their long, crocodilian beaks they gobbled up strange fish whose fossils we have since found in mud that turned to stone. At night, these terrifying creatures, called pteranodons or pterodactyls, clung to cliffs with the claws on their wings, in the same way that bats sleep today.

For a long time, it was thought these flying reptiles were the ancestors of modern birds, but as paleontolo-

They have conquered the planet

gists gradually unraveled the tangled skein of evolution, they discovered this was not so. These terrifying creatures, the pterosaurians, in fact have no descendants. So, what were the first birds, then?

Imagine another flashback, this time to about 150 million years ago, to a worldwide Jurassic Park reigned over by *Tyrannosaurus rex.* Here, another winged creature, a lot less ugly than the pterosaur, glided from tree to tree. Its body was covered with feathers and, if the fossils unearthed at the end of the nineteenth century are anything to go by, it did indeed resemble a kind of bird with a long tail. Could this creature, the archaeopteryx, have been the long-lost ancestor? Paleontologists had their reservations. For with its beak lined with teeth, its three long, clawed fingers protruding from each wing, and its tail like a lizard's, this bizarre animal seemed

to have been more reptile than bird. Disappointed, they put the fossils discovered in Bavaria back into their specimen drawers.

Later, in the United States, another fossil was discovered, one that fit the bill, far better. This was a flying vertebrate the size of a crow, probably covered with feathers, and with shoulders and forelimbs prefiguring those of birds. Could the *Protoavis texensis* (it was discovered in Texas) have been the missing link? The jury is still out, but most specialists agree it is the oldest "bird" in the world.

Later, as fossils discovered in Arkansas show, birds more worthy of the name progressively appeared on Earth in various species—some, like the icthyornis (fish eaters), even resembled present-day gulls, only with teeth. Others, the hesperornis, were more akin to the divers we know today, except they were more than 3 feet tall. It would seem like the mystery of birds' distant ancestry had finally been solved. But in fact, none of these discoveries enabled paleontologists to crack the enigma of the missing link. That is, what creatures, at a given point in evolution, enabled the crossover from reptile to bird?

Why do seabirds never get wet?

In order to repeatedly dive and swim underwater, whether in freshwater or in the sea, aquatic birds must never allow their plumage to soak up water, as doing so would make them heavier and affect their performance. So how do their feathers remain watertight? The answer is that their uropygial gland, the oil-secreting gland every bird has above its tail, is far more developed. Waterbirds collect the "grease" from the gland on their beak or bill and meticulously spread it over their body. Cormorants are the only birds lacking this gland, hence the "crucified" pose they adopt to dry their plumage in the sun or wind as they perch on branches after fishing.

Wandering albatross

All hypotheses lead back to the dinosaurs, some of whom, certain specialists believe, after long evolution developed the ability to glide. Their forelimbs gradually transformed into wings, their jaws into a beak, and their scales into feathers. The dinosaurs that became bipeds about 100 million years ago, such as the 10-foot-long velociraptor, so called because it ran very fast, were the ancestors of the theropods, giant creatures covered with feathers and possessing a skeleton enabling flight. Some paleontologists believe the theropods were the ancestors of birds, and they did indeed have the prerequisite equipment for flight: fused clavicles and a lunatum in each wrist that enabled them to move them sideways. Still, the mystery is far from being solved: we know dinosaurs could not fly because they were far too heavy.

Does this mean the "dinosaur trail" is a dead end? The recent discovery of fossils of tiny theropods in China has given new heart to its strongest advocates. Dubbed microraptors, these "pocket dinosaurs" were capable of flying—or, rather, they could have given birth to a lineage that itself evolved into the warm-blooded vertebrates we call birds. At any rate, the Chinese microraptors have taught us a lot about the evolution of feathers.

The one thing we are sure of today is that birds descended from reptiles. They both share certain particularities in their nervous, respiratory, and circulatory systems and bird feathers are merely evolved scales. Before leaving this quest for the avian grail, however, consider the most

Black-browed albatross,
Falkland Islands.

incredible hypothesis evolutionists have yet concocted: birds' closest present-day relatives are . . . crocodiles!

What exactly do we qualify as a "bird"? According to authorities, a bird is an oviparous vertebrate covered with feathers, with pulmonary respiration and warm blood. It uses its hind legs for walking and its forelimbs or wings for flying (except for apterous birds such as ostriches and nandus, or common rheas, which are unable to fly).

If we abandon the quest for the mysterious common ancestor, we can nevertheless ask ourselves, at what moment did true birds—that is, creatures with all the characteristics stipulated by the above definition—really appear on Earth? Tens of thousands of years ago? Millions of years ago? Tens of millions of years ago? We must leap back into the past again to meet feathered creatures resembling those we know today. More than 60 million years ago, during the Cretaceous period, lived the first creatures worthy of the name "birds." These were wading birds, rails, and cormorants—easily comparable to their present-day namesakes.

Millions of years later, the bird family diversified with geese, ducks, gannets, curlews, moorhens, birds of prey, and owls, all of which flew, laid eggs, and brooded exactly like birds do today. If a contemporary ornithologist could step into a time machine and travel 35 million years back into the past, he would be right at home with the company he saw: large grouse, pelicans, vultures, penguins,

Wandering albatross, Crozet Island.

High-fliers

A part from vultures, particularly Ruppel's vultures, one of whom collided with a passenger plane at 36,000 feet, the highest flyers in the world are:

- *Bar-headed geese, which on their twice-yearly migration over the Himalayas, between China and India, fly at altitudes of around 30,000 feet.*
- *Whooper swans, some of which have been spotted at altitudes of more than 26,000 feet between Iceland and England.*
- *Common swifts, which sometimes reach that altitude following clouds of insects carried skyward by hot air pockets.*
- *Black-tailed godwits, which can reach around 20,000 feet.*
- *Storks, which fly at up to 16,000 feet.*
- *Lapwings, which fly at up to 13,000 feet, as well as thrushes (11,000 feet), American swans (5,000 feet), and chaffinches (6,000 feet).*
- *At any lower altitude, most birds can be encountered primarily passerines, which account for more than half of all bird species.*

puffins, flamingos, ducks, pewits, owls, sparrows, and a host of passerines almost identical to those we know today. And if this ornithologist decided to hop 10 million years forward toward the present, he would feel even more at home—or at least as if he were in tropical Africa—in the forests of the Miocene age, teeming with pelicans, marabous, ibises, and parrots.

Birds, therefore, have gone a long time without changing their appearance—far longer than any mammal. When *Homo sapiens,* a hairy mammal among so many others, appeared on Earth, birds had already been flying for millions of years.

There is no place on Earth birds have not progressively colonized, no region on land or sea, from the Arctic to Antarctica, in the Old World and the New World, that is out of the reach of their wings and that they do not continue to frequent today. Inhabiting torrid deserts, icebergs, soaring peaks, desert islands, cities, virgin forests, idyllic Edens, and seemingly uninhabitable hells, birds are everywhere. They are the only vertebrates to have conquered the entire planet, the only creatures to have so efficiently made light of earthly contingencies—first and foremost gravity.

Round-the-world yachtsman Yves Parlier, fascinated by the spectacle of a wandering albatross escorting his fragile boat through a storm in the South Seas, wrote in his logbook, "You whose voice I have never heard, I suppose you must owe your name to the territory you are the lord of, the Furious Fifties. I am totally fascinated by you.

Black-browed albatross in the Roaring Forties, Falkland Islands.

How did nature create such a pure marvel? I dream of being able to rid myself of gravity, to roam the ocean from wave to wave like you. You are a privileged being among animals; your realm has never changed, its extent has never diminished, and you enjoy the same freedom as you did centuries ago."

The legendary albatross, Baudelaire's giant "prince of the clouds," is one of *Winged Migration*'s most impressive stars. It has a 13-foot wingspan and (along with the Andean condor) is one of the world's biggest birds, but what makes this peerless glider even more extraordinary is that it is capable of traveling thousands of miles while using hardly any energy. And it does so out of choice: the albatross, the pelagic bird par excellence, spends more than 90 percent of its life at sea, flying alone over the ocean, only returning home to one of those tiny specks of terra firma in the southern oceans to reproduce.

Winged Migration's chief wildlife camera operator, Thierry Thomas, and a three-person film crew stayed on Crozet Island for several months. He noted in his shooting journal, "In gusts, the wind must easily reach 100 mph. The water of waterfalls is blown away without reaching the ground. We attempted a short trip to the albatross field, bent double, leaning at a 45-degree angle into the wind. And there the birds were, firmly ensconced on the nests, all facing into the wind, totally indifferent to the weather." Thomas's daunting description speaks worlds much about the stoicism of this noble bird.

Black-browed albatross waiting for an updraft to lift it into the air, Falkland Islands.

Winged water carriers

The males of species of sandgrouse living in arid habitats in the Old World frequently cover distances of up to about 60 miles to drink at permanent water sources. Having filled up, the sandgrouse shakes itself in the water to dampen its plumage, particularly its belly, which is covered with a special kind of down. Having thus transformed itself into airborne water carriers, it wings its way back to the nest, where it lets its thirsty chicks suck water from its feathers.

In Africa, turtledoves in the Sahel regions also make daily trips to water sources during the nesting season. They store water in their crop, which they then regurgitate for their chicks when they return to the nest.

The home of these lords of the extreme is the territory of the Roaring Forties and the Furious Fifties, the volatile waters located roughly in the area between 40 degrees and 60 degrees latitude south, where none except a few lone yachtsmen dare to venture. Yet the albatross seems perfectly content to live in this oceanic apocalypse, perfectly in its element in the howling winds and raging seas. In fact, nothing could suit it better. In order

to glide just above the waves, these great silver birds rely on the wind—the stronger the better. Their nightmare is fair weather, which brings their majestic flight to a standstill. "Grounded," they bob up and down on the water like vulgar sea-gulls, waiting for the wind to return.

Yannick Clerquin, one of Galatée Films' ornithologists, who also stayed on Crozet Island, recalls: "When there's a depression over the austral seas, it's as if there's a black wall on the horizon, and in front of that wall you make out flecks of light flitting in all directions, albatrosses celebrat-ing the approaching storm."

Thanks to the Argos satellite tracking sys-tem, we now know that these wandering birds are capable of covering distances of several thousand miles at speeds of up to 50 mph by simply allow-ing themselves to be carried along on the west winds. In a single year, they can circumnavigate the globe several times without touching the ground. Their solitary voyages over the immensity of the southern oceans can last several years, until decked out in its splendid nuptial plumage, young albatrosses return to their birthplace, one of the tiny specks of land floating in the immense sea. There, on Crozet or another island, it will spread its wings and perform an incredible dance, turn-ing around and tapping its beak to its partner's in a sort of kiss.

The incubation period of the albatross's single egg, three months, is the longest among all birds. The chick is fed on the fish and squid "soup" its parents regurgitate for it on their return

Wandering albatross sleeping with its beak under its wing, Crozet Island.

Black-browed albatross looking down over its nesting colony, Falkland Islands.

from one of their long-distance fishing trips. Chicks can go for a week without eating, although albatross infant mortality is very high, with almost 60 percent never making it to adulthood. On the other hand, their life expectancy is the highest of any bird, 95 percent living to the age of fifty or even sixty. Their longevity is unique among birds, and so is their fidelity. Albatross couples remain together for life, despite not seeing each other for months at a time each year.

Another living legend is the penguin. Scientists have recently solved some of the mysteries concerning these birds' exceptional aptitude for adapting to the extreme conditions of the South Atlantic. A team of French, German, and British ornithologists studied the aquatic performance of king penguins (slightly smaller than emperor penguins), which can dive to a depth of about 1,000 feet for up to eight minutes. How do they manage this feat, unthinkable for humans, without any apparent effort, scores of times a day during the fishing season? Researchers have discovered that they deliberately cool certain parts of their body so as to burn less energy during dives. This adaptation allows them to go into localized hypothermia and is unique in the animal kingdom. Certain mammals such as dormice, marmots, and hedgehogs are capable of self-induced hypothermia, but it is a generalized state sending them into deep sleep. The penguin's system, however, is infinitely more subtle. Other birds, too, can voluntarily go into a kind of hibernation. In

King penguin with its egg on its webbed feet, Falkland Islands.

extremely cold weather, for example, Andean hummingbirds shelter in caves, where their body temperature can descend from 104°F to 66°F.

French researchers from the Center of Ecology and Energetic Physiology, in Strasbourg, followed the parental odyssey of the king-penguin colony on Crozet Island, filmed by the *Winged Migration* team. These birds can swim up to 400 miles to go fishing, but it quite often happens—four times out of ten—that a female does not return to land until after her egg, left in her partner's care, has hatched. In this case, the chick may for the first few days have been fed solely by its father, which himself has not eaten for several weeks. Fed with what, then? A team of French scientists led by Professor Yvon Le Maho, a leading specialist in penguins, discovered that male birds deliberately retain food in their stomach in order to regurgitate it into the beak of their chick if the mother is late returning from fishing. Even more surprising is that a father penguin, himself living on energy reserves, prevents his metabolism from digesting these emergency provisions. And how he manages to keep the food fresh at 100°F in his stomach is another as-yet-unsolved mystery.

Contrary to what one might think, only 11 percent of bird species have a "lifestyle" involving an aquatic environment. There are in fact two types of seabirds: shorebirds, such as seagulls, that rarely stray far from coasts, and deep-sea birds, such as petrels, puffins, and albatrosses, lords of the high seas and stars of *Winged Migration.*

King penguin sitting on its egg, which is protected by a fold of skin on its belly, Falkland Islands.

Gentoo penguin calling its mate,
Falkland Islands.

Gentoo penguin, Falkland Islands.

Birds can live in all climates, from the storms of the North Atlantic to the blizzards of the Antarctic. Few human beings have ventured into the icy, windswept wastes around the South Pole, one of the most inhospitable regions of the world, but those who have have been amazed to find snow petrels there—the only living creatures to survive in such conditions.

There are also birds in the hottest places in the world, in scorching savannas, steppes, and deserts in which the only other inhabitants are a few reptiles, insects, or small mammals. The Australian parrot, the Saharan lark, the sandgrouse of Central Asia, the roadrunner of the American west, the turtledove of South Africa, and the Arabian sparrow hawk have all conquered the most inhospitable deserts. Certain species live in regions where the ground temperature can reach 158°F. To do so, they must conquer dehydration. Some drink cactus sap, others morning dew; birds of prey rely on the digestive juices of their catches (lizards and small rodents); others, such as ducks and pink flamingos, resort to nomadism, traveling alone or in groups to regions where it has just rained. It is a mystery just how these pilgrims know rain has fallen hundreds of miles away. Perhaps they have a kind of meteorological sixth sense shared with a few desert mammals such as the famous addax antelope. For as soon as a carpet of green springs up, these nomadic birds waste no time, immediately

parading, coupling, and then building a nest as if there were no tomorrow.

Another thing that differentiates desert species from other birds is that many do not have an annual reproduction cycle. Instead, they reproduce whenever and wherever the opportunity arises—that is, when it rains, an event they may have to wait months, even years, for. Some, such as Californian bobwhites, lay only when rain comes, and never in drought years. The longer the period of enforced sterility, the more eggs they produce.

Another specificity of desert birds is that they have light plumage when they live in areas where the ground is light colored and darker plumage when the terrain is dark, camouflage being their prime means of escaping the attention of birds of prey, foxes, and jackals. But ornithologists have observed that plumage color also varies according to ambient humidity: the higher the humidity level, as in tropical zones, the more brightly colored the bird's feathers, the lower the humidity, the quieter the plumage. The plumage of various members of the same species, therefore, varies according to habitat.

* *

*

Storms, snow and ice, deserts, oceans: birds have conquered them all. They have also defeated unimaginably high altitude—the rarified, freezing upper layers of the atmosphere where oxygen is scarce. This is the altitude at which you might

A moment of tenderness between a rockhopper penguin and its chick, Falkland Islands.

Rockhopper penguins, Falkland Islands.

cross the Himalayas in a plane. It's the altitude at which the last thing you would expect to see outside your window is a bar-headed goose—a flying machine weighing about 4 to 6 pounds and capable of cruising at more than 30,000 feet in conditions of −122°F. In fact, a plane once collided with a Ruppel's vulture at more than 35,000 feet. Nobody would have believed the pilot's account in the plane's logbook had not the traces of the impact also been recorded after the plane landed. Why? Because the collision occurred at an altitude at which, theoretically, no living thing could survive.

No living thing, except a bird.

Sipping saltwater

Birds are the only vertebrates, except for tortoises and the Galapagos marine iguana, capable of living for long periods in a hypersaline environment. So, how to they drink? Biologists initially thought birds had a specific system of renal elimination, but this was proved not to be the case. Intrigued by the small glands seabirds have above their eye sockets, lodged in short grooves—an anatomic feature no other birds have—ornithologists focused their research on them: the secretions that flow from these "nostrils," which in the albatross are very deep, contain 5 percent salt, 2 percent more than seawater. These glands are covered with cells that literally suck the sea salt out of the bird's blood, acting like physiological filters. Equipped with these organic dialysis machines, seabirds can evacuate the surplus salt they absorb from their food.

Rockhopper penguins on their way to the sea, Falkland Islands.

Rockhopper penguin, Falkland Islands.

Pages 194–195: Anhinga drying its wings, Bharatpur National Park, India.

Birds in danger

by Guy Jarry, assistant director of the Center for Research on the Biology of Bird Populations at the Muséum national d'histoire naturelle (National Museum of Natural History), Paris

At least 200 million years ago, when the scales on the body of a small dinosaur were replaced by the first primitive forms of feathers, what might have seemed to be a strange mutation soon proved to be advantageous to the creatures who found themselves covered with these appendages, since not only did they provide better insulation but they also put the animals on the evolutionary track leading to birds as we know them today. The regular discovery of new species of dinosaur, whose more or less intact remains have been found fossilized in rock or sediment, has helped build up a more complete picture of the extraordinary origins of birds.

Down through the ages, along the long march of evolution, feathered dinosaurs multiplied and diversified. The enormous biped dinosaurs discovered in South America had forelimbs that were still clawed but so thickly feathered it seems they were used as short wings to improve balance while running. More dinosaur fossils with forelimbs covered with remiges not yet enabling them to fly have been discovered in China. However, the jury is still out as to the origins of flight. Were these small, feathered dinosaurs runners, treehoppers, or jumpers? Did they live on the edges of lakes and marshes, capable of swimming, running on nearby land, and climbing rocks and trees? In such habitats, they could have benefited from specific aerological conditions (types, frequency, and intensity of winds) that may have enabled them to acquire the ability to glide and then progress to wing-beating flight.

About 130 million years ago, there lived a creature called the archaeopteryx, considered by many to be the first bird, combining reptilian characteristics (well-developed teeth, wings equipped with fingers and claws, and a long, feathered tail similar to that of the biped dinosaurs) and avian characteristics (body covered with feathers, forelimbs already transformed in wings). The first true birds, anatomically comparable to today's species, appeared a few million years later.

Between the end of the Tertiary (Miocene) and the Quaternary Periods—that is, over a period from 26 million to 10,000 years ago—present-day species appeared and progressively diversified over the entire surface of our planet, adapting to varying favorable climates and increasingly diverse habitats. There are now less than 10,000 bird species, present on virtually every piece of land protruding from the sea. They constitute the culmination of a 130 million-year process, the tip of an evolutionary iceberg of more than 150,000 now-extinct bird species. Yet, even at the end of the Tertiary era, in the Miocene and Pliocene periods (6 million to 2 million years ago), considered to be the time frames most favorable to bird diversity, the number of species at any given moment would not have exceeded 32,300.

While new species obeying evolution's complex mechanisms slowly emerged (and are still doing so), others became extinct, either progressively or suddenly. The many causes of extinction included failure to adapt to sudden climate and habitat changes, predation by large mammals, epidemics, natural catastrophes such as volcanic eruptions and tidal waves caused by earthquakes or violent storms, the direct action of cyclones and hurricanes, and the devastating impacts of giant meteorites. The species most exposed to these risks were those with small populations confined to limited and vulnerable geographic areas (low islands, the flanks of volcanoes, or rare habitats). This remains the case today.

But there is one cause of extinction perhaps more redoubtable than any other: humanity. Already in prehistoric times, humans are thought to have largely contributed to the disappearance of a great many species, particularly on islands. Humankind, now omnipresent on the planet's surface with its ever-growing population, is having an increasingly devastating effect on the quality and abundance of Earth's natural biodiversity. Whereas the average rate of extinction of wild bird species should not exceed 1 species per century, 128 species have disappeared since 1500, and 103 have become extinct since 1800—twenty-five times greater than the natural rate of extinction—largely due to humanity's ravages or to indirect effects of it.

The disappearance of most species has been brought about by the immoderate hunting, capturing, and sometimes even extermination of a bird population for food, for commercial reasons, or for sale as pets. In many cases, the destruction of a specific and exclusive habit, or the fortuitous or deliberate introduction of predatory mammals—a phenomenon particularly common on tropical islands—has brought their demise. Excessive hunting has killed off once-flourishing populations of migratory pigeons and the Carolina parakeet in North America, the Cuban macaw in the Antilles, the great auk on the islands of the North Atlantic, the black emu in Australia and the Mascarene Islands, the dodo on Mauritius and

Reunion Island, and the solitaire on Rodrigues Island, to cite but a few well-known examples.

The disappearance of many other species, such as parrots on many islands in the Antilles and in the Indian Ocean and the Pacific Ocean, was principally due to the annihilation of their habitats. The situation is exacerbated by excessive persecution and hunting, with the traffic for the pet-bird market playing a preponderant role.

A Birdlife International report on endangered bird species, published in 2000, listed 1,186 species, 12 percent of Earth's avifauna, threatened worldwide according to the criteria established by the World Conservation Union (UICN). Three species, the red-billed curassow of Brazil, the Socorro (Grayson's) dove of Mexico, and the Guam rail, disappeared recently, and 182 other species, such as the Cuban ivory-billed woodpecker, the Eskimo curlew in North America, Spix's macaw in South America, the slender-billed curlew in Asia, and the bald ibis in Morocco and the Near East are on the verge of extinction. Some of these gravely threatened species now number only a few dozen birds, or at the most a few hundred, and extinction of another 321 species may be imminent due to their dangerously low populations. Still another 680 species are considered to be highly vulnerable, risking extinction in the medium term, while 727 others, due to their low populations, require constant surveillance. To which one should add the 959 species with a breeding habitat of no more than 30,000 square miles and the 470 species with populations of less than 2,500. If humanity's devastating behavior continues at its present rate, we have every

Macaws on a clay cliff, Peru.

reason to fear that during the first century of the third millennium, another 400 species will disappear forever.

The categories of main threats humans pose for the entire animal and vegetable kingdom are the alteration or, more radically, destruction of habitats. This includes the excessive exploitation of tropical-forest resources, immoderate hunting and trapping, various kinds of pollution, and the increasingly perceptible alteration of the climate due to global warming, caused in particular by pervasive combustion of fossilized carbon fuels (coal, gas, and oil). Birds fall victim to all these threats as well as other accidental ones. Tropical regions are among those most affected by these problems, whether in Asia (Indonesia, China, India, the Philippines) or South America (Brazil, Colombia, Peru, Ecuador), the countries that rank highest in the number of bird species threatened with extinction on their territory.

Europe is equally as concerned by these problems. Of the 513 species comprising its nesting avifauna, 195 (38 percent) have unfavorable preservation prospects, their populations being either menaced, in decline, or vulnerable. In Europe, as elsewhere, the degradation of habitats and of the environment in general, as well as hunting and persecution, are the principal threats.

France is especially concerned by the threatened extinction of birds. Of the 324 species regularly nesting or wintering there, 155 are classified as disfavored or fragile, and the spotted eagle, the lesser kestrel, the corncrake, and the aquatic warbler are also on the list of endangered species worldwide. But the French state also shares responsibility for another

54 endangered species that live in its overseas departments and territories.

The drawing up of such a precise planetary picture has involved countless and often costly studies and research projects, many of which have been carried out in difficult conditions and have included the wholesale mobilization of both amateur and professional ornithologists. The evaluation of the conservation status of a species, an essential stage in our knowledge of it, necessitates accurately identifying the type and importance of the factors determining it.

The planning of appropriate conservation measures would be impossible without these preliminary scientific studies. Scientists and conservationists are mobilizing themselves, sometimes in a headlong race, to come to the assistance of the most endangered species, which increasingly frequently have to be put "on a drip" if they are to be saved. But all too frequently, alas, such efforts come up against powerful economic interests, political reticence, the hostility of social and professional lobbies, the spinelessness of states, or, even worse, the deliberate will to do nothing. Faced with man's increasing pillage of nature, one sometimes wonders whether the struggle to save our planet's natural biodiversity, daily becoming an ever-greater challenge, is in vain. A species' extinction is an irremediable tragedy for nature and, for science, but it is also a symptom of the degradation of our environment. As such, it should be taken as a powerful alarm signal because it is also a direct effect of the growing threats that loom over the future of humankind.

Protecting birds

*M*any bird species are endangered, but certain birds are in demographic growth again due to the protection they now benefit from. Such is the case of the snow goose, whose hunting had to be authorized again in Quebec to keep the population down. The most edifying example, through, is the white stork in Alsace, whose population today is twice as large as it was in 1900.

Traditionally, the white stork has always wintered in Mali, Nigeria, and Senegal, but for several years now, more and more have been spending the winter in Morocco, Spain, and even southern France. The first white storks return to nest in Europe around mid-February, the males arriving first, making the journey from Mali to Alsace or its new nesting zones in Normandy and on the Atlantic coast in about twenty days.

In 1947, there were only 177 couples in France and in 1974, only 9, and their extinction seemed inevitable. The situation was the same in all the countries of western Europe. But effective protection measures were taken in the 1980s, with repopulating ensured by birds raised in captivity, and the trend reversed. In 1990, 138 couples returned to nest in France, then 430 in 1997, 180 of them in Alsace and the rest along the Atlantic and Channel coasts. Where there's the will and a way: with effort, even endangered birds can be saved.

White stork.

Certain characteristics have enabled birds to colonize the entire planet. What distinguishes them foremost from other animals is their feathers and, with a few exceptions, their ability to fly. The second important point, shared with mammals, is their maintenance of a constant body temperature of between 100° and 104° Fahrenheit.

Flight enables numerous species to live in regions whose seasonal temperature differences can be enormous. These species have had to acquire specific aptitudes to survive in inhospitable habitats, often covered with deep snow, such as the Arctic and Antarctic. Grouse, for example, dig burrows in the snow in which to spend the night. Other birds hoard food in caches during the summer months, and are able to find them again in winter under as much as 3 feet of snow. The only species to reproduce in midwinter, the emperor penguin, does so in Antarctica. Whereas all other polar species reproduce during the summer months, the emperor penguin waits until the sea has begun to freeze over. The males incubate the egg in midwinter, in the polar night, in temperatures of about −100° Fahrenheit and in winds sometimes exceeding 125 mph.

Birds have conquered the polar zones, but they are also common in desert zones, where, yet again, their extraordinary ability to adapt enables them to survive, with some species able to reach extremely distant water sources. Several species also encounter extreme situations during migration, such as the bar-headed goose, which crosses the Himalayas at altitudes of up to 30,000 feet on its yearly migration to India.

How can birds fly at such high altitudes, in temperatures so low? Lack of oxygen is the main problem, since even more energy is needed to fly at these altitudes. But birds have a more high-performance respiratory system than mammals and have a comparatively larger heart. These dual characteristics enable more rapid and efficient oxygenation of body tissues. The very low temperatures encountered at 25,000 to 33,000 feet by migratory birds often used to tropical temperatures are therefore not a problem; on the contrary they are an advantage. Intense muscular effort generates excess heat, which is dissipated more easily by the very cold air, which also reduces the dehydration this muscular effort causes.

Even more surprising, though, is birds' extremely efficient exploitation of the marine environment. They are to be found on every sea in the world, often at considerable distances from the nearest continent or island. For the most part, these are migratory birds crossing stretches of water where they cannot feed and where many cannot land. Many limicolous birds cross the Pacific Ocean from Alaska to New Zealand, for example, stopping on only a few islands.

Marine birds spend most of their lives out at sea. This is where they sleep, molt, and find their food, returning to land only to reproduce. Frigate birds, for example, never alight on the ocean surface, and some believe they sleep on the wing like swifts.

Although most seabirds mainly exploit the sea's surface, some can dive to considerable depths in search of food: the petrel and the sooty shearwater to more than 200 feet, for example, and cormorants and guillemots to between 400 and 500 feet. But the greatest divers of them all are the penguins. These birds, which have lost the ability to fly, can reach depths of more than 1600 feet, and the emperor penguin is capable of remaining underwater for more than twenty minutes. How, then, can air-breathing birds with lungs transform themselves into fish? Several specific adaptations enable them to do so. Their body is shaped like a torpedo, and they propel themselves with wings that have evolved into highly efficient swimming fins—so efficient that to watch a penguin swimming underwater is like watching one flying.

Other physiological adaptations enable penguins to stay at great depth for considerable lengths of time. They store oxygen in their muscles, ten times more than other birds, while their respiratory sacs contain enough oxygen for the effort of the dive. Penguins can also reduce their internal temperature, another means of economizing on oxygen.

These extraordinary examples demonstrate how, thanks to flight and a host of other morphological, behavioral, and physiological adaptations, birds, like mammals, have colonized the entire planet.

Polar temperatures, lack of oxygen— nothing bothers them

by Henri Weimerskirch CNRS research director at the Center for Biological Studies Chizé

Bar-headed geese in the Himalayas.

Pages 202–203: White storks in Alsace, France.

The cast of *Winged Migration*

"I think I could turn and live with animals, they are so placid and self-contain'd
I stand and look at them long and long.
They do not sweat and whine about their condition,
They do not lie awake in the dark and weep for their sins,
They do not make me sick discussing their duty to God,
Not one is dissatisfied, not one is demented with the mania of owning things,
Not one kneels to another, nor to his kind that lived thousands of years ago,
Not one is respectable or industrious over the whole earth."

Walt Whitman

From Northern Europe to Africa

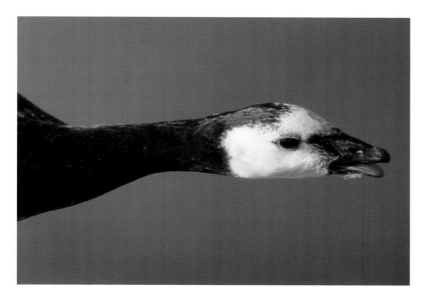

BARNACLE
GOOSE
Branta leucopsis

Length: 23 to 31 in.
Wingspan: 51 to 57 in.
Weight: 2.9 to 4.4 lb.
Flight speed: 38 to 50 mph
Characteristics: A small, pretty goose nesting in loose colonies on cliffs, safe from predators such as the arctic fox. It is herbivorous, with precocial chicks. Highly sociable in winter, which it spends in large, noisy flocks on damp coastal grasslands. So-called because in the Middle Ages it was thought to be hatched from barnacles. At the end of reproduction it molts completely, which prevents it from flying for three to four weeks.

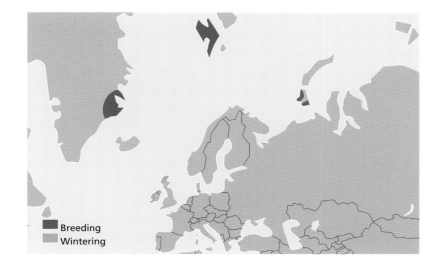

GREYLAG
GOOSE
Anser anser

Length: 29 to 35 in.
Wingspan: 58 to 70 in.
Weight: 5.5 to 9.7 lb.
Flight speed: 37 to 50 mph
Characteristics: The most common gray goose and the ancestor of most domestic geese. Flies in V formation during migration. It is herbivorous, with precocial chicks. Frequents estuaries, lakes, coastal wetlands, and pastures. Aggressive on its breeding territory, it becomes very gregarious in winter and migrates in large flocks. At the end of the breeding period, it molts completely, which prevents it from flying for three to four weeks.

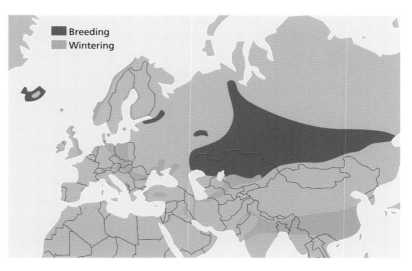

Breeding
Wintering

COMMON
CRANE
Grus grus

Length: 43 to 47 in.

Wingspan: 86 to 96 in.

Weight: 11.2 to 13.4 lb.

Flight speed: 28 to 43 mph

Characteristics: The common crane, one of the largest European birds, lives in the great humid forests of northern Europe and Russia. It is omnivorous, with precocial chicks. The courtship display is a spectacular dance accompanied with calls. Solitary during reproduction, it migrates in large groups toward Spain. During winter, it feeds in open spaces, often in cultivated areas. Its characteristic call during migration carries a long way.

ARCTIC
TERN
Sterna paradisaea

Length: 14 in.

Wingspan: 30 to 33 in.

Weight: 3.4 to 5.1 oz.

Flight speed: 25 mph

Characteristics: One of the farthest-migrating birds and a superb and elegant long-distance flyer due to its slim, aerodynamic body, lengthened by two long tail streamers. It nests in colonies on the ground in open areas along coastlines. Highly territorial and aggressive toward intruders. Piscivorous, the arctic tern hovers high over the water before swooping on its prey. Before and after nesting, the male often brings its mate small fish.

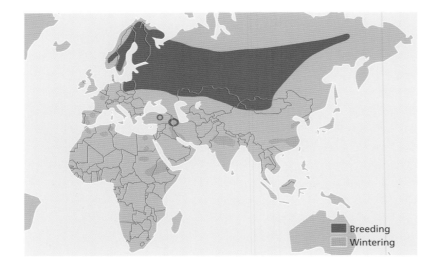

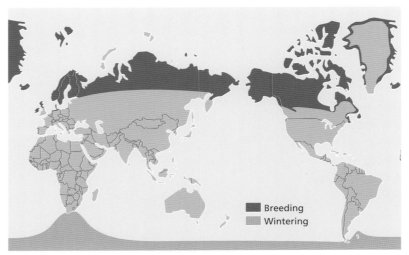

From Northern Europe to Africa

WHITE
PELICAN
Pelecanus onocrotalus

Length: 55 to 68 in.

Wingspan: 91 to 140 in.

Weight: 20 to 33 lb.

Flight speed: 19 to 31 mph

Characteristics: One of the largest flying birds in the world, the white pelican has to run on the water to take off. Ascends by gliding on thermals but is also capable of sustained flapping flight. Flies with its neck curled back to better support the weight of its enormous beak. It is piscivorous, with altricial chicks, and nests in colonies. White pelicans fish collectively in a semicircle, driving fish toward the middle, where they use their beak pouch—which has a capacity of around 2 gallons—like a landing net.

WHITE
STORK
Ciconia ciconia

Length: 39 to 45 in.

Wingspan: 68 to 76 in.

Weight: 5.1 to 9.7 lb.

Flight speed: 28 mph

Characteristics: A large gliding bird that migrates only by day, since it rides thermals created by the heat of the sun. Characteristic black-and-white plumage, bright red beak and feet. It is omnivorous, with altricial chicks that remain in the nest until they have attained adult size and can fly. Nearly always frequents open spaces, damp or dry. Builds its nest on top of a tree or structure. Generally mute, they greet each other at the nest by loudly snapping their beaks together.

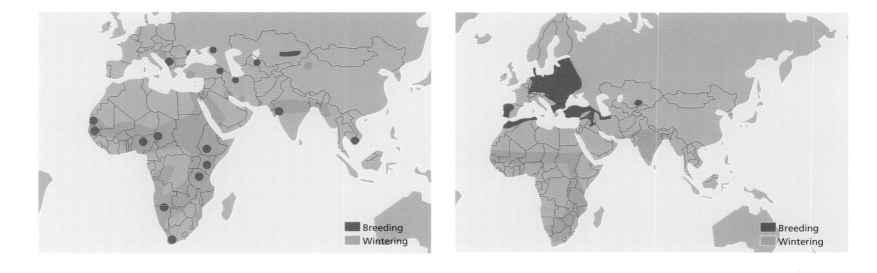

Breeding
Wintering

Breeding
Wintering

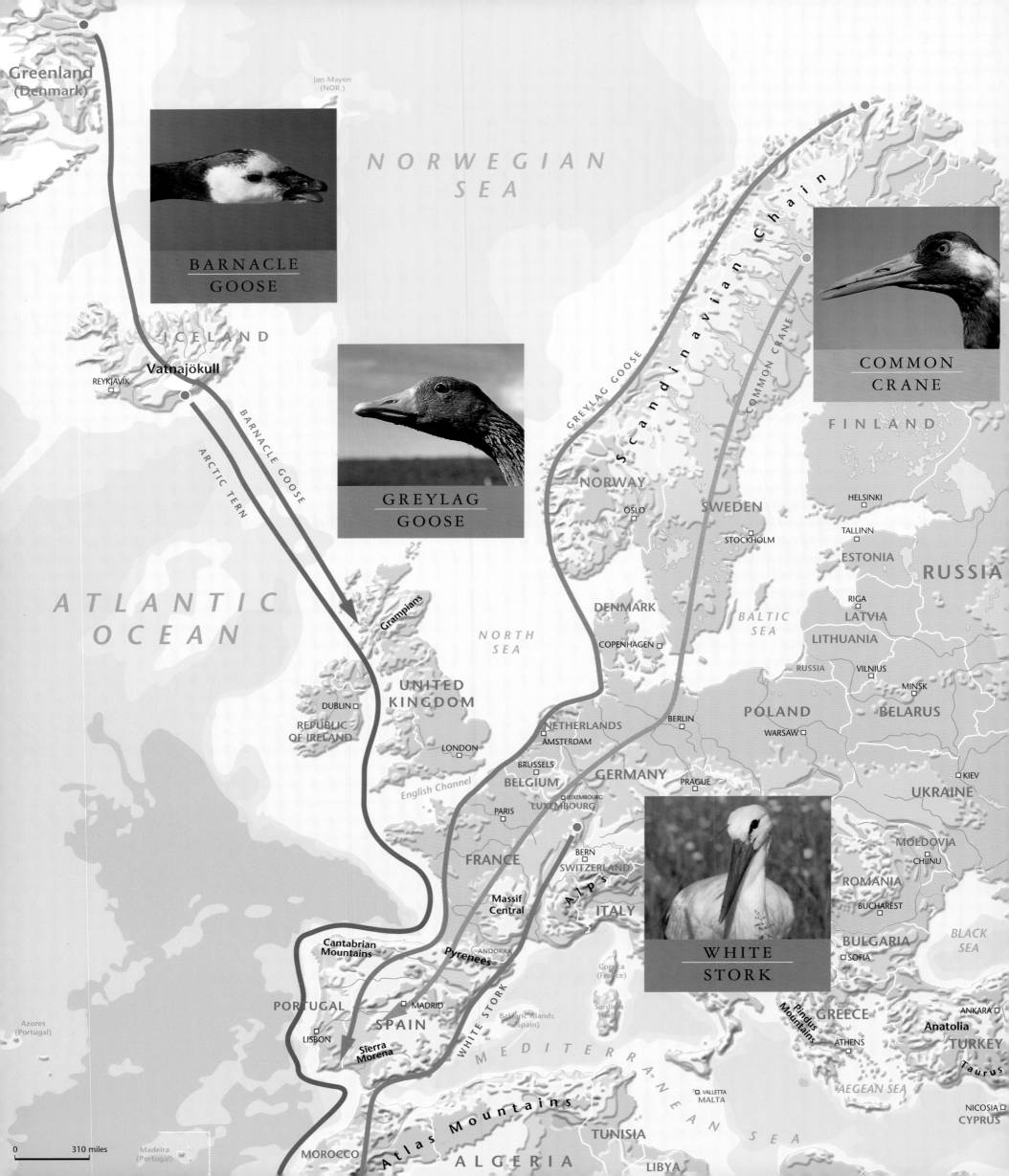

Greenland
(Denmark)

NORWEGIAN
SEA

Jan Mayen
(NOR.)

BARNACLE
GOOSE

ICELAND

Vatnajökull

REYKJAVÍK

BARNACLE GOOSE

ARCTIC TERN

ATLANTIC
OCEAN

GREYLAG
GOOSE

Scandinavian Chain

GREYLAG GOOSE

COMMON CRANE

COMMON
CRANE

FINLAND

NORWAY

OSLO

SWEDEN

STOCKHOLM

HELSINKI

TALLINN

ESTONIA

RUSSIA

DENMARK

COPENHAGEN

BALTIC
SEA

RIGA

LATVIA

LITHUANIA

Grampians

NORTH
SEA

RUSSIA

VILNIUS

MINSK

BELARUS

DUBLIN

UNITED
KINGDOM

LONDON

REPUBLIC
OF IRELAND

English Channel

PARIS

NETHERLANDS

AMSTERDAM

BERLIN

POLAND

WARSAW

BRUSSELS

BELGIUM

GERMANY

PRAGUE

KIEV

UKRAINE

LUXEMBOURG

LUXEMBOURG

FRANCE

BERN

SWITZERLAND

Massif
Central

Alps

ITALY

MOLDOVIA

CHIINU

ROMANIA

BUCHAREST

WHITE
STORK

Cantabrian
Mountains

Pyrenees

ANDORRA

Corsica
(France)

BULGARIA

SOFIA

BLACK
SEA

PORTUGAL

MADRID

Balearic Islands
(Spain)

Sardinia
(Italy)

Pindus
Mountains

GREECE

Anatolia

ANKARA

LISBON

SPAIN

Sierra
Morena

WHITE STORK

MEDITERRANEAN

ATHENS

TURKEY

Taurus

Azores
(Portugal)

AEGEAN
SEA

NICOSIA

CYPRUS

VALLETTA

MALTA

SEA

Atlas Mountains

Madeira
(Portugal)

MOROCCO

ALGERIA

TUNISIA

LIBYA

0   310 miles

ATLANTIC
OCEAN

PARIS

BERN

VIENNA BRATISLAVA

Carpathian Mtn.

MOLDOVA

KAZAKHSTAN

FRANCE

SWITZERLAND

AUSTRIA HUNGARY BUDAPEST

CHIŞINĂU

Massif
Central

Alps

Po

SLOVENIA

LJUBLJANA ZAGREB

ROMANIA

ITALY

CROATIE

BELGRADE BUCHAREST

AZOV
SEA

BOSNIA-
HERZEGOVINA

SARAJEVO

YUGOSLAVIA

Balkan Mtn.

Danube

BULGARIA

BLACK SEA

GEORGIA

Caucasus

TBILISI

CASPIAN
SEA

Cantabrian
Mountains

Pyrenees

ROME

Apennines

ALBANIA MACEDONIA

TIRANA

SKOPJE SOFIA

Pindus
Mountains

ARMENIA
YEREVAN

AZERBAIJAN

BAKU

ANDORRA

GREECE

ATHENS

ANKARA

TURKEY

Lake
Van

Kuro

Zagros Mtn.

PORTUGAL

MADRID

SPAIN

Balearic Islands
(Spain)

ALGIERS

TUNIS

MALTA

VALLETTA

AEGEAN SEA

Anatolia

Taurus Mtn.

Lake
Tuz

Iris

Tigris

Lake
Urmia

TEHRAN

LISBON

Azores
(Portugal)

WHITE STORK

Atlas Mtn.

TUNISIA

NICOSIA

CYPRUS

LEBANON

SYRIA

BEIRUT

DAMASCUS

JERUSALEM

ISRAEL

AMMAN

JORDAN

Dead Sea

BAGHDAD

IRAK

Euphrates

IRAN

Hejaz

RABAT

MOROCCO

TRIPOLI

Madeira
(Portugal)

CAIRO

Syrian
Desert

KUWAIT

KUWAIT

Tropic of Cancer

WESTERN
SAHARA

ALGERIA

Libyan Desert

EGYPT

SAUDI
ARABIA

RIYADH

MANAMA

BAHRAIN

QATAR

DOHA

UNITED
ARAB
EMIRAT

Canary Islands
(Spain)

MAURITANIA

Tibesti Mtn.

WHITE
PELICAN

Nile

Nubian
Desert

Rub' al-Khali

OMAN

CAPE VERDE
ISLANDS

NOUAKCHOTT

Senegal

CHAD

SUDAN

ERITREA

Asir

YEMEN

Hadramavt

DAKAR

SENEGAL

Niger

MALI

KHARTOUM

ASMARA

SANAA

BANJUL

GAMBIA

BAMAKO

N'DJAMENA

DARFUR

DJIBOUTI

DJIBOUTI

BISSAU

GUINEA-
BISSAU

GUINEA

Black Volta

Lake
Volta

NIGERIA

Chari

Nil Blanc

Lake
Tana

ADDIS ABABA

CONAKRY

FREETOWN

SIERRA
LEONE

IVORY
COAST

Niger

Benue

Adamawa

Ethiopian
Plateau

Ogaden

ETHIOPIA

MONROVIA

GHANA

LOME

ABUJA

CAMEROON

CENTRAL
AFRICAN REPUBLIC

Nil Bleu

SOMALIA

YAMOUSSOUKRO

TOGO

BENIN

PORTO-NOVO

LIBERIA

ACCRA

Bioko
(Eq. G.)

BANGUI

Uele

MOGADISHU

Equator

GULF OF GUINEA

SÃO TOMÉ
& PRÍNCIPE

YAOUNDÉ

Ubangi

Congo

UGANDA

KENYA

ARCTIC
TERN

LIBREVILLE

EQUATORIAL
GUINEA

REPUBLIC
OF THE
CONGO

DEMOCRATIC
REPUBLIC OF THE
CONGO

RWANDA

KAMPALA

Lake
Victoria

KIGALI

NAIROBI

Tana

GABON

Congo

BUJUMBURA

BURUNDI

TANZANIA

Lualaba

Galana

SEYCHELLES

BRAZZAVILLE

Kasai

DODOMA

Rufiji

KINSHASA

Cuango

Lake
Tanganyika

LUANDA

Cuanza

Lake Nyasa

Ruvuma

MORONI

COMOROS

Bié
Plateau

MALAWI

Luiio

ANGOLA

ZAMBIA

LILONGWE

MADAGASCAR

LUSAKA

Zambeze

ATLANTIC
OCEAN

Kwango

HARARE

ANTANANARIVO

Namib Desert

NAMIBIA

ZIMBABWE

MOZAMBIQUE

BOTSWANA

Save

Tropic of Capricorn

WINDHOEK

Indian Ocean

ARCTIC TERN

GABORONE

PRETORIA

MAPUTO

Kalahari
Desert

MBABANE

SWAZILAND

Orange

MASERU

Vaal

REPUBLIC OF
SOUTH AFRICA

LESOTHO

CAPE TOWN

Great Karoo

Crozet Islands
(France)

620 miles

Scale at the Equator

North America

BALD
EAGLE
Haliaeetus leucocephalus

Length: 28 to 37 in.
Wingspan: 66 to 95 in.
Weight: 6.6 to 14 lb.
Flight speed: 19 to 31 mph
Characteristics: This fishing eagle, national emblem of the United States, snatches its prey from the water's surface with its talons. Can be necrophagous. Lives beside large stretches of water, lakes, or the sea. During courtship, the male and female sometimes lock talons and free-fall, twirling as they plummet toward the ground, then parting at the very last moment.

SNOW
GOOSE
Anser caerulescens

Length: 28 to 30 in.
Wingspan: 51 to 63 in.
Weight: 6.6 to 11 lb.
Flight speed: 34 to 59 mph
Characteristics: The snow goose is easily recognizable by its white plumage highlighted by black wing tips and pinkish beak. Extremely gregarious during wintering. It is herbivorous, with precocial chicks. The snow-goose population is rapidly growing because during winter, it feeds on crops in agricultural regions of the United States and southern Canada and because it is a protected species. At the end of the breeding period, it molts completely, which prevents it from flying for three to four weeks.

CANADA
GOOSE
Branta canadensis

Length: 21 to 43 in.
Wingspan: 47 to 74 in.
Weight: 4.4 to 15.4 lb.
Flight speed: 43 to 56 mph
Characteristics: North America's most familiar and most common goose. Frequents a variety of habitats ranging from the tundra to inland semideserts and forests, all of which have one constant: nearby water. Quebecois call it the honker goose because of the loud call it makes during flight. Can be almost as large as a swan. It is herbivorous, with precocial chicks. Very close-knit families; the parents are extremely aggressive around the nest. Gregarious during wintering. At the end of the breeding period, it molts completely, which prevents it from flying for three to four weeks.

Breeding
Wintering

Breeding
Wintering

Sedentary flocks
Breeding
Wintering

BEAUFORT SEA

Banks
Island

Greenland
(Denmark)

Victoria
Island

BAFFIN BAY

Baffin Island

Arctic Circle

Mackenzie Mountains

Mackenzie

Great Bear
Lake

Melville
Peninsula

ATLANTIC
OCEAN

LABRADOR
SEA

Yukon

Rocky Mountains

Coastal Ranges

Great Slave
Lake

SNOW GOOSE

BALD EAGLE

Peace

Lake Athabasca

Great Plains

C A N A D A

Labrador

Fraser

Athabasca

Saskatchewan

Lake
Winnipeg

Hudson Bay

Quebec

Columbia

Nelson

CANADA GOOSE

Newfoundland

Cascade Range

Rocky Mountains

Missouri

Lake Superior

Saint Lawrence

Columbia

Great
Basin

Snake

Great Plains

Mississippi

Lake Michigan

Lake
Huron

Lake
Ontario

OTTAWA

Coastal Ranges

Sierra Nevada

Great Salt
Lake

U N I T E D   S T A T E S   O F   A M E R I C A

Colorado

Rio Grande

Missouri

Lake Erie

Ohio

WASHINGTON

Appalachians

Tennessee

Red

ATLANTIC

Baja California

Western Sierra Madre

Rio Grande

Mississippi

OCEAN

Florida

Tropic of Cancer

PACIFIC

MEXICO

GULF OF
MEXICO

NASSAU

BAHAMAS

OCEAN

MEXICO CITY

Balsas

Southern Sierra Madre

HAVANA

CUBA

DOMINICAN
REPUBLIC

Yucatan

BELIZE

BELMOPAN

GUATEMALA

GUATEMALA CITY

HONDURAS

SAN SALVADOR

TEGUCIGALPA

SALVADOR

NICARAGUA

MANAGUA

JAMAICA

KINGSTON

HAITI

PORT-AU-
PRINCE

SANTO
DOMINGO

Puerto Rico
(USA)

CARIBBEAN
SEA

ANTIGUA AND
BARBUDA (U.S.A.)

SAINT KITTS
AND NEVIS

Guadeloupe (Fr.)

DOMINICA

Martinique (Fr.)

SAINT LUCIA

BARBADOS

SAINT VINCENT AND THE GRENADINES

GRENADA

TRINIDAD AND TOBAGO

0   620 miles

South America

ANDEAN CONDOR
Vultur gryphus

Length: 39 to 51 in.
Wingspan: over 125 in.
Weight: 17 to 33 lb.
Flight speed: 12 to 19 mph
Characteristics: One of the world's largest gliding birds, along with pelicans and albatrosses. Entirely black except for white wing markings. Lives in mountainous regions in South America, but can be found at low altitude in Peru, Chile, and Patagonia, where it sometimes frequents the coastline. An exclusively carrion-eating bird, it feeds in groups. Nests in colonies in cliff crevices.

BLUE AND GOLD MACAW
Ara ararauna

Length: 34 in.
Wingspan: 28 to 30 in.
Weight: 2.2 to 2.9 lb.
Flight speed: 22 to 43 mph
Characteristics: An Amazonian forest dweller feeding exclusively on extremely hard seeds, which it chisels into with its powerful beak. The blue and gold macaw nests in a hole hollowed out of a palm trunk. Its call can be heard from a great distance. Couples remain together for life. In certain regions, macaws gather on cliffs to eat clay, which counteracts the toxins in the grains they feed on.

■ Distribution area

■ Distribution area

ROCKHOPPER
PENGUIN
Eudyptes chrysocome

Length: 21 to 24 in.

Fins: 6 to 7 in.

Weight: 4.8 to 8.8 lb.

Swimming speed: 4.3 to 5 mph

Characteristics: Like all penguins, the rockhopper has a second, transparent eyelid to protect the eye underwater. Nests on rocky or steep terrain, jumping from rock to rock to return to its colony, which is dense and often shared with other species. Couples recognize each other by distinctive calls. It is piscivorous, and the chicks gather together in small crèches. Feathers are specially adapted for life in cold water and provide insulation.

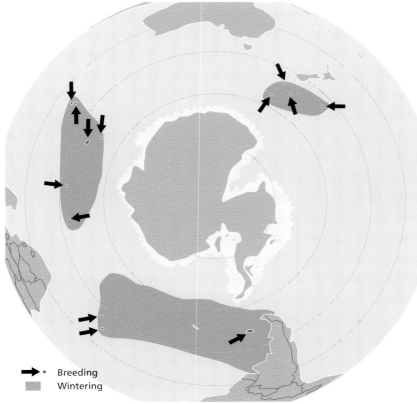

➤• Breeding

Wintering

MANAGUA

Lake
Nicaragua

SAN JOSÉ

COSTA RICA

PANAMA

PANAMA

CARACAS

PORT-OF-SPAIN

GRENADA

TRINIDAD AND
TOBAGO

VENEZUELA

GEORGETOWN

PARAMARIBO

GUYANA

SURINAME

FRENCH
GUIANA

Magdalena

Meta

Llanos

Orinoco

BOGOTÁ

Guaviare

Branco

Galápagos Islands
(Equator)

Equator

QUITO

COLOMBIA

ECUADOR

Putumayo

Japurá

Negro

Amazon

Amazon

A T

Andes

Ucayali

Juruá

Purus

Tapajós

Xingu

BRAZIL

LIMA

PERU

BLUE AND GOLD MACAW

Araguaia

Tocantins

São Francisco

Espinhaço Mountains

Lake
Titicaca

LA PAZ

BOLIVIA

Planatto do
Mato Grosso

Serra Geral

BRASÍLIA

Altiplano

Uyuni
Salt Flat

ANDEAN
CONDOR

Atacama
Desert

Tropic of Cancer

Atacama Desert

Paraguai

Paraná

Chaco

ASUNCIÓN

Gran

Rio Bermejo

PARAGUAY

Serra do Mar

Paraná

Uruguay

PACIFIC

OCEAN

Andes

Grande
Salt Flat

URUGUAY

Uruguay

Paraná

MONTEVIDEO

SANTIAGO

Pampas

BUENOS
AIRES

ARGENTINA

Rio Colorado

Rio Negro

CHILE

ANDEAN CONDOR

Patagonia

Falkland Islands (U.K.)

Tierra
del Fuego

0   620 miles

Scale at the Equator

GUINEA
BISSAU
GUINEA
MALI
OUAGADOUGOU
NIGERIA
TCHAD
CONAKRY
SIERRA
LEONE
IVORY
COAST
Black Volta
Lake
Volta
TOGO
BENIN
Niger
ABUJA
Benue
Chari
N'DJAMENA
FREETOWN
GHANA
LOMÉ
CENTRAL AFRICAN REPUBLIC
MONROVIA
YAMOUSSOUKRO
ACCRA
PORTO-
NOVO
CAMEROON
Adamawa
BANGUI
LIBERIA
Bioko (Equatorial Guinea)
YAOUNDÉ
Uele
SÃO TOMÉ AND
PRÍNCIPE
EQUATORIAL
GUINEA
LIBREVILLE
Ubangi
Congo
REPUBLIC OF THE CONGO
GABON
Congo
DEMOCRATIC
REPUBLIC
OF THE CONGO
BRAZZAVILLE
Kasai
Lualaba
KINSHASA
Kwango
LUANDA
Cuanza
Bié Plateau
ANGOLA
ZAMBIA
LUSAKA
Okavango
Zambezi
Lake
Kariba
ZIMBABW
Namib Desert
NAMIBIA
BOTSWANA
WINDHOEK
GABORONE
PRETORIA
Kalahari
Desert
Vaal
MASERU
ATLANTIC
Orange
OCEAN
REPUBLIC OF
SOUTH AFRICA
CAPE TOWN
Great Karoo

NTIC
OCEAN

BLUE AND GOLD
MACAW

ROCKHOPPER
PENGUIN

ROCKHOPPER PENGUIN

Georgia (U.K.)

Sandwich Islands (U.K.)

Asia

RED-BREASTED
GOOSE
Branta ruficollis

Length: 20.5 to 21.5 in.

Wingspan: 45 to 53 in.

Weight: 2.2 to 3.3 lb.

Flight speed: 50 to 62 mph

Characteristics: A small, very fast-flying goose with characteristic black, red, and white plumage. Nests in dry tundra near the nest of a large bird of prey so as to benefit from its protection against predators. It is herbivorous, with precocial chicks. Highly gregarious during winter. At the end of the breeding period, it molts completely, which prevents it from flying for three to four weeks.

BAR-HEADED
GOOSE
Anser indicus

Length: 28 to 30 in.

Wingspan: 55 to 62 in.

Weight: 4.4 to 7 lb.

Flight speed: 37 to 50 mph

Characteristics: A medium-size goose with elegant plumage living on the high plateaus of Central Asia, always near water. Forms dense colonies. It is herbivorous, with precocial chicks. The bar-headed goose is one of the highest-flying species: to reach its winter quarters, it has to fly over the highest mountains in the Himalayas (over 30,000 ft). At the end of the breeding period, it molts completely and cannot fly for three to four weeks.

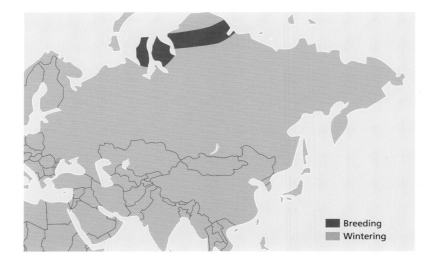

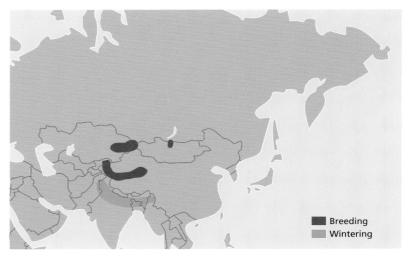

BLACK-HEADED
IBIS
Threskiornis melanocephalus

Length: 25 to 30 in.

Wingspan: 43 to 47 in.

Weight: 3.3 lb.

Flight speed: 20 to 30 mph

Characteristics: Lives in humid areas scattered with trees, on one of which it builds its nest in colonies of varying density. It is omnivorous. Migratory flight either in V formation or in rank. The head is completely bald, revealing its black skin.

JAPANESE
CRANE
Grus japonensis

Length: 58 in.

Wingspan: 86 to 97 in.

Weight: 15 to 26 lb.

Flight speed: 30 to 50 mph

Characteristics: An omnivorous species endangered by destruction of its winter habitats. On the ground, the Japanese crane folds its wings in such a way that the black remiges form a kind of plume above the white rump. The courtship display is a dance interspersed with frequent jumps and accompanied by calls.

WHOOPER
SWAN
Cygnus cygnus

Length: 55 to 64 in.

Wingspan: 84 to 96 in.

Weight: 17 to 29 lb.

Flight speed: 35 to 50 mph

Characteristics: One of the largest flying European birds, the whooper swan has more than 25,000 feathers. The male aggressively defends the nest against all intruders. It is herbivorous, with precocial chicks. Gregarious during wintering. Its peculiar call gave rise to the legend of the swan song. The whooper swan has to run on the water to take off.

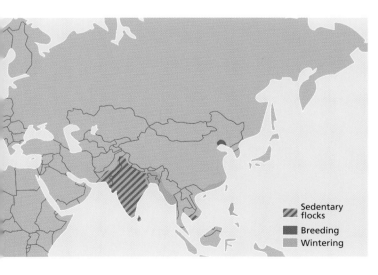

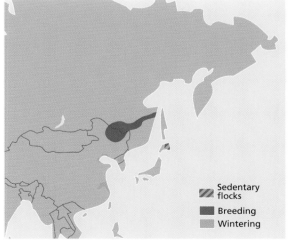

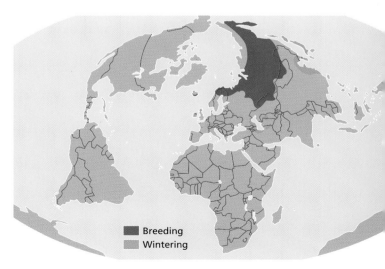

BARENTS
SEA

KARA
SEA

Taymyr Penin

Pjasina

RED-BREASTED GOOSE

NORWAY

Kjølen Mountains

Kola
Peninsula

Yamal
Peninsula

Yenisey

SWEDEN

Arctic Circle

Pechora

East

China

R

Lower Tunguska

FINLAND

*Lake
Onega*

*Lake
Ladoga*

□ HELSINKI

U
r
a
l

M
o
u
n
t
a
i
n
s

Ob'

Plains

Stony Tunguska

□ STOCKHOLM

BALTIC SEA

□ TALLINN

ESTONIA

RED-BREASTED
GOOSE

Irtysh

Yenisey

□ RIGA

LATVIA

Tobol

□ MOSCOW

RUSSIA

BAR-HEADED
GOOSE

*Hövsö
Nuu*

LITHUANIA

□ VILNIUS

□ MINSK

Ishim

□ AQMOLA

POLAND

BELARUS

Volga

□ WARSAW

Vistula

Oder

Pripet

KAZAKHSTAN

*Lake
Zaysan*

*Hyargas
Nuur*

Mongolian Altai Nuruu

□ KIEV

UKRAINE

Don

Gob

Carpathian Mtn.

SLOVAKIA
□ BRATISLAVA
□ BUDAPEST
HUNGARY

MOLDOVA
□ CHIŞINĂU

Caspian Depression

*Ustyurt
Plateau*

M

□ ZAGREB

ROMANIA

BAR-HEADED GOOSE

□ BELGRADE
BOSNIA-
HERZEGOVINA
□ SARAJEVO
YUGOSLAVIA
KOSOVO
□ SOFIA
□ TIRANA
MACEDONIA
ALBANIA

□ BUCHAREST

Danube

BLACK SEA

Caucasus Mtn.

CASPIAN SEA

UZBEKISTAN

*Lake
Balkhach*

T
i
a
n

S
h
a
n

Lop Nur

Qinling Shandi

Qin

GEORGIA
□ TBILISI

□ BAKU

□ TASHKENT

KYRGYZSTAN

Kaxgar

Altun Shan

ATHENS □

GREECE

ANKARA □

TURKEY

YEREVAN □
ARMENIA
AZERBAIJAN

*Lake
Van*

TURKMENISTAN

TAJIKISTAN
□ DUSHANBE

Kunlun Shan

C H I

Tigris

□ TEHRAN

□ ASHKHABAD

Tibet

MEDITERRANEAN SEA

NICOSIA □

CYPRUS

Euphrates

SYRIA

Plateau of Iran

□ KABUL

H

i

m

a

l

a

y

a

s

T I B E T

BEIRUT □
LEBANON
ISRAEL
□ JERUSALEM

□ DAMASCUS

IRAQ

I R A N

Indus

ISLAMABAD □

AFGHANISTAN

□ AMMAN

□ BAGHDAD

JORDAN

KUWAIT

□ KUWAIT CITY

PAKISTAN

Sutlej

NEPAL

BHUTAN
□ THIMPHU

Brahmaputra

Salween

CAIRO □

A
r
a
b
i
a

□ DELHI

BHUTAN
□ THIMPHU

LIBYA

*Libyan
Desert*

EGYPT

Nile

RED SEA

BAHRAIN
QATAR
□ RIYADH   □ DOHA
UNITED ARAB
EMIRATES
□ ABU
DHABI

Indus

KATHMANDU □

I N D I A

Ganges

BANGLADESH

□ DACCA

Brahmaputra

Me

Tropic of Cancer

SAUDI
ARABIA

□ MUSCAT

Narmada

MYANMAR
(BURMA)

*Tibesti
Mountains*

*Nubian
Desert*

Rub al-Khali

OMAN

ARABIAN SEA

Mahanadi

Godavari

YANGON □

VIENTIA

CHAD

□ KHARTOUM

SUDAN

ERITREA
□ ASMARA

□ SANAA

YEMEN

White Nile

Blue Nile

Krishna

BAY
OF
BENGAL

0   620 miles

Scale at the Equator

EAST
SIBERIAN
SEA

LAPTEV
SEA

*Taymyr
Lake*

Olenek

Lena

Yana

Indigirka

Indigirka

Verkhoyanskiy Khrebet

Vilyui

Lena

Vilyui

Kolyma Mountains

Kolyma

Anadyr

SEA
OF
OKHOTSK

BERING
SEA

Kamchatka

Amur

WHOOPER SWAN

**WHOOPER
SWAN**

Lena

Vitim

*Lake
Baikal*

Yablonovyy Range

**JAPANESE
CRANE**

JAPANESE CRANE

Manchuria

Hulun Nur

AANBAATAR

Kerulen

OLIA

sert

*Lake
Khanka*

NORTH
KOREA

PYONGYANG

SEOUL

SOUTH
KOREA

JAPAN

TOKYO

PACIFIC

OCEAN

BEIJING

Huang

Huang

NA

Chang

*Dongting
Hu*

*Poyang
Hu*

YELLOW
SEA

EAST
CHINA SEA

SEA
OF JAPAN

BLACK-HEADED IBIS

TAIPEI

TAIWAN

**BLACK-HEADED
IBIS**

NOI

VIETNAM

SOUTH
CHINA SEA

MANILA

PHILIPPINES

DIA

Antarctic circle

KING
PENGUIN
Aptenodytes patagonicus

Length: 37 in.
Fins: 12 to 14 in.
Weight: 20 to 33 lb.
Swimming speed: 7.5 mph
Characteristics: One of the two largest penguins. Like all penguins, it cannot fly but is perfectly adapted to swimming underwater. It is piscivorous and lays only one egg, which it incubates on its feet to protect it from the cold. Lives in very dense colonies of up to a million birds. The chicks gather in huge crèches to wait for their parents to return from fishing.

WANDERING
ALBATROSS
Diomedia exulans

Length: 43 to 53 in.
Wingspan: 100 to 135 in.
Weight: 13 to 24 lb.
Flight speed: 25 to 50 mph
Characteristics: One of the largest seabirds and a supreme long-distance flyer, the albatross covers vast distances using variations in air currents above the waves. A carrion eater, it feeds on floating dead fish and cephalopods. Couples remain together for life. During the courtship display, the male and female spread their wings, snapping their beaks and calling. The period from laying the egg to the chick taking flight lasts almost a year.

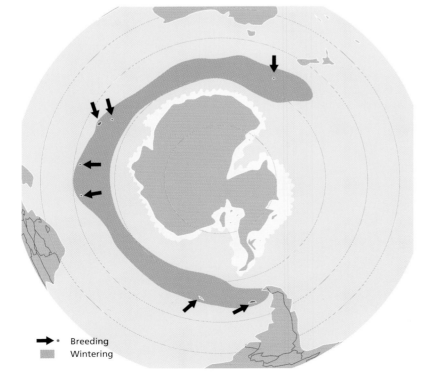

➤• Breeding
  Wintering

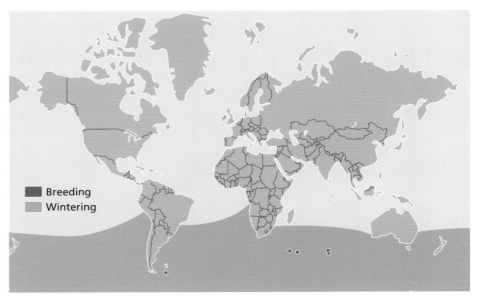

■ Breeding
■ Wintering

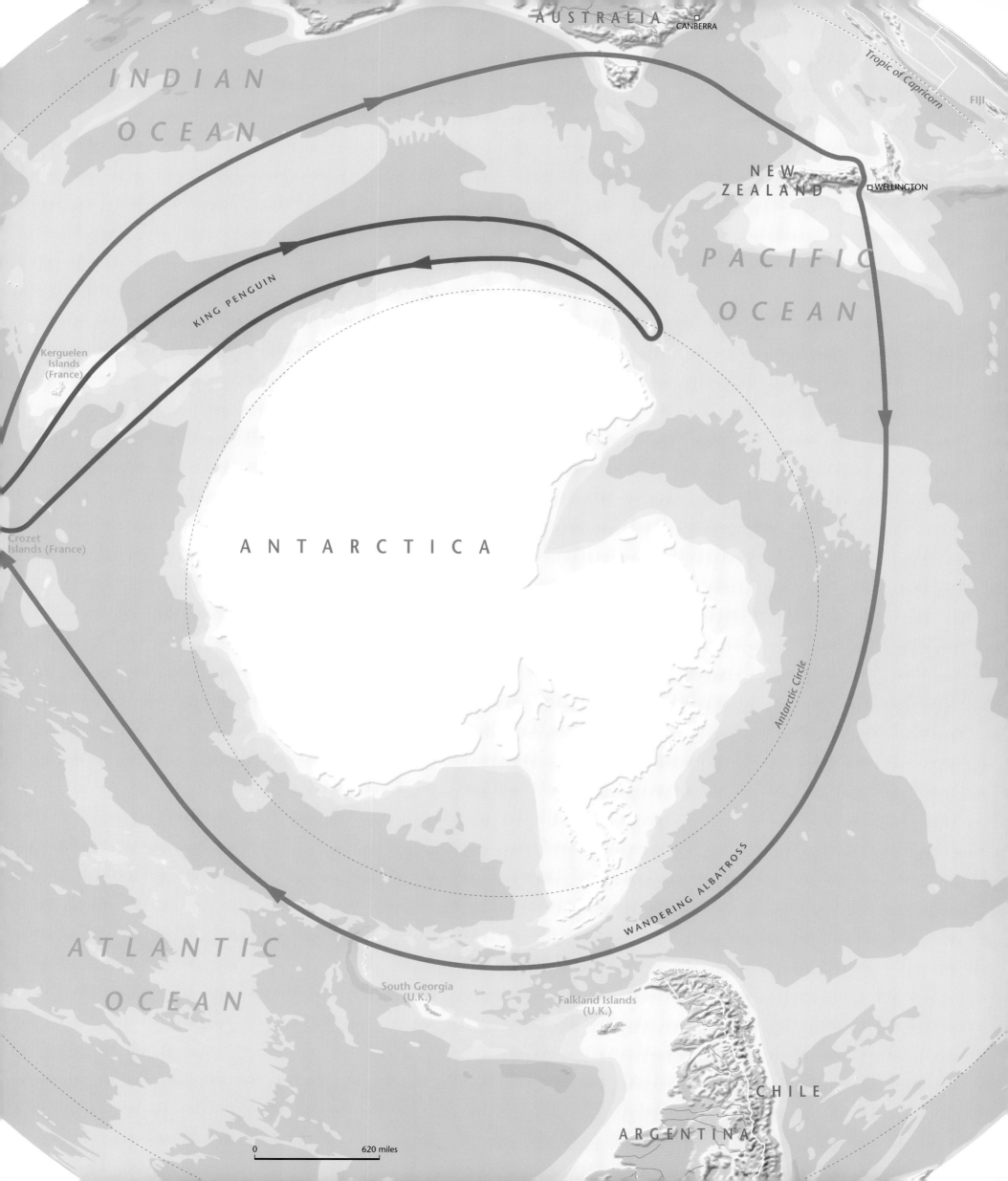

AUSTRALIA
□ CANBERRA

INDIAN

OCEAN

Tropic of Capricorn

FIJI

NEW
ZEALAND
□ WELLINGTON

PACIFIC

OCEAN

KING PENGUIN

Kerguelen
Islands
(France)

Crozet
Islands (France)

ANTARCTICA

Antarctic Circle

WANDERING ALBATROSS

ATLANTIC

OCEAN

South Georgia
(U.K.)

Falkland Islands
(U.K.)

CHILE

ARGENTINA

0   620 miles

Michel Debats

Jacques Perrin

Jacques Cluzaud

BEHIND THE SCENES

Arizona

Arizona

New York

Adirondacks

▲ Paris ▼

▼ Jura, France

▲ Shooting in Monument Valley, Utah/Arizona.

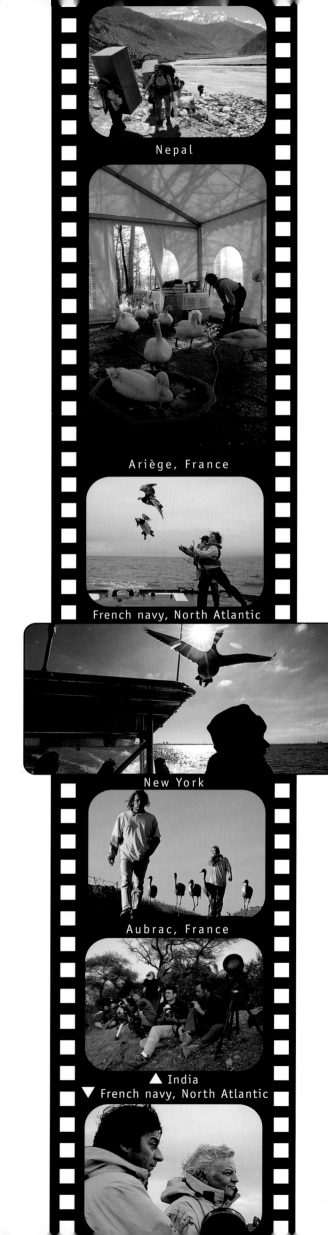

Nepal

Ariège, France

French navy, North Atlantic

New York

Aubrac, France

▲ India
▼ French navy, North Atlantic

Who has never dreamed of joining one of those great undulating flocks of geese flying over our concrete jungles?

Who has never dreamed of gliding beneath the stars on silken wings, of discovering the wonders of the great wide world anew, with the eyes of a bird?

Who, on a summer night, has never longed to streak through the blue on great beating wings, whirling and squawking with a flight of swifts, drunk with freedom?

Who, in a word, has never dreamed of flying with the birds, of ridding oneself of gravity and roaming through the air wherever one wills?

To fly with the birds, to physically and psychologically become part of their winged migration, to fly around the planet from Asia to Australia, from Canada to Argentina, from Europe to Africa, from the Arctic to the Antarctic, rid at last of all human frontiers. To be a bird, if only for an hour, a single hour . . . .

This dream has haunted humanity since the dawn of time. Even the very recent conquest of the air by flying machines, only a century ago, has only partially banished the thought from our imagination.

The story of *Winged Migration* is just that dream.

It all began in 1997, when Jacques Perrin, producer of *The Monkey People* (1988) and *Microcosmos* (1996), decided to complete his

▲ Training geese in Normandy, France.

▼ Preparing to shoot on Skrudur Island, Iceland.

trilogy with a new cinematic adventure, this time one devoted to birds. "I wanted this film to be a great natural fable," he explains, "a fable drawing its power solely from its imagery. When we see a little half-ounce bird in our back yard, we don't realize it has come 6,000 miles, braving storms and a thousand dangers to be there. I wanted us to rediscover our planet through birds."

This crazy idea had haunted Jacques Perrin since boyhood, when he used to marvel at the cranes flying over his boarding school. The kid from Joinville-le-Pont with his head in the clouds grew up, but he never forgot those migrating birds. It was these supremely free creatures he wanted to get close to, to film like never before. This feat would take years of continuous shooting in some forty countries, with 300 miles of film, 150 staff and crew members, dozens of ornithologists, and a colossal budget invested by Jacques Perrin's production company, Galatée Films, to achieve.

"Three directors—Michel Debats, Jacques Cluzaud, and myself, and eight co-directors, in fact camera operators—can all share credit for this film," Jacques Perrin continues. "At the bottom of our hearts, we all wanted to do the same thing: to pay tribute to birds. What we wanted, above all, was to get close to their mystery."

And, to approach that mystery on the surest footing, the producer, now turned director, called on some of the world's greatest ornithologists. The late Professor Jean Dorst, a specialist in South American birds and former director of the

National Museum of Natural History in Paris, the late Professor Francis Roux, an expert on West African birds and previously assistant director of the Laboratory of Mammals and Birds at the National Museum of Natural History, Guy Jarry, a specialist in the birds of the regions between Europe and West Africa and deputy director of the Center for Research on the Biology of Bird Populations at the National Museum of Natural History, all endorsed the project.

Numerous foreign specialists also gave their support, such as Professor Kenneth P. Able of New York State University, a specialist in migratory navigation, Peter Berthold at the Max-Planck Institute Vogelwarte in Germany, a specialist in the migration of warblers and storks, Dr. Hiroyoshi Higuchi of Tokyo University, a specialist in Asian cranes, and Dr. Yossi Leshem of the Department of Zoology at Tel Aviv University, a specialist in pelicans, birds of prey, and storks. All these great minds came together to form the film's scientific committee, a genuine council of wise men that received the support of international organizations as prestigious as Birdlife International, the French League for the Protection of Birds, the Audubon Society in America, the National Museum of Natural History, and the one million–member Royal Society for the Protection of Birds in the United Kingdom.

At the same time, *Winged Migration*'s fledgling team was also joined full time by two young French ornithologists, Stéphane Durand

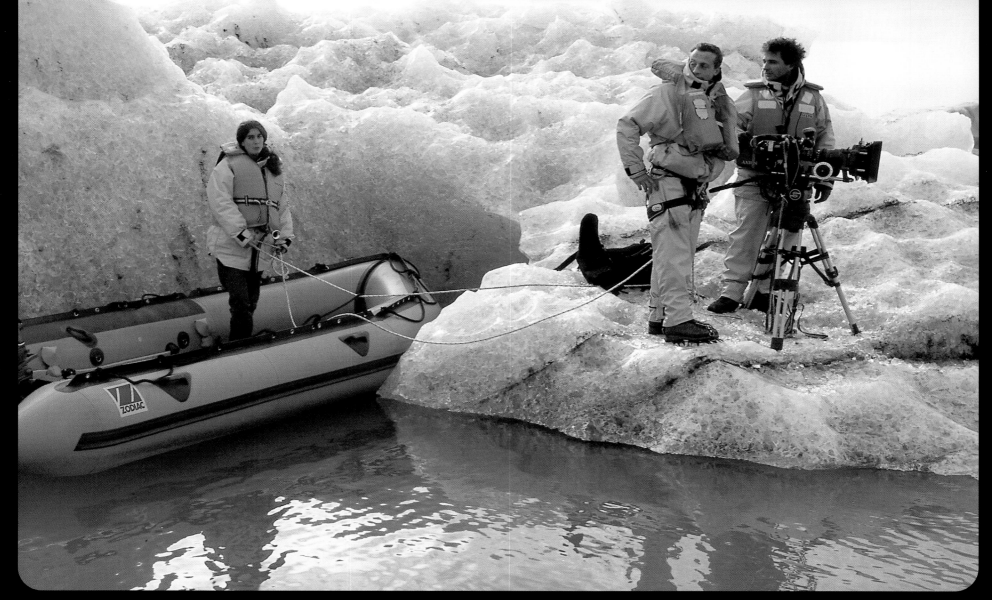

▲ Shooting in a glacial lagoon, Iceland.   ▼ Shooting on the Hudson River, New York.

TRAVELING SHOT ON THE HUDSON RIVER,
ALTITUDE 1000 FEET.

END OF FLIGHT AROUND THE STATUE OF LIBERTY.
THE GEESE LEAVE THE FRAME.

and Guillaume Poyet, and all these experts worked together for a year to develop Jacques Perrin's and Stéphane Durand's script treatment.

These specialists, for the most part field researchers, indicated to the film team the most interesting nesting and wintering locations in the world, often in rarely seen and very difficult-to-reach places such as Crozet Island in the South Atlantic, visited only twice a year by a French naval ship, and Skrudur Island off the coast of Iceland, a privately owned island hosting huge colonies of puffins, guillemots, and gannets. Some of these eminent ornithologists did not hesitate to take part in location scouting and shooting expeditions. Francis Roux, for example, left his work for several weeks to supervise the extensive filming on the Arguin Banks off the coast of Mauritania.

On a daily basis, month after month, all these experts kept a watchful eye on the scientific rigor of this ambitious venture, one that had to both adhere to reality and sublimate it, one so exalting in the extraordinary experiences it created that it progressively brought together some ninety researchers from all over the world—from Argentina, Brazil, Germany, India, Japan, Russia, Sweden, and the United States—to form one of the largest "private" ornithological networks the world has ever known.

"We made a film that stuck as close as possible to the reality of birds," says Laurent Fleutot, *Winged Migration*'s director of photography. The idea wasn't to steal pictures, as wildlife camera operators usually do, but to fly in the company of birds."

The magic of *Winged Migration* lies in its harmonious blend of these two approaches, that of the wildlife filmmaker and that of the traditional filmmaker.

To achieve this close-up intimacy and to film continuity and link footage for the editing, birds had to be raised, trained, and made accustomed to humans in captivity. The techniques employed were perfected with wild geese in the 1930s by Austrian naturalist Konrad Lorenz, considered today to be the father of ethology (the science of animal behavior). Pictures of this white-bearded scientist with goslings perched on his head are ubiquitous, and many know of his experiments, which involved taking care of young geese in order to become their veritable adoptive father. He was the first to understand that even before hatching, from the beginning of incubation, one had to stroke the eggs and talk to them softly through the shell to accustom them to one's voice and human presence.

Despite being branded a crackpot, the future Nobel Prize winner persisted with his strange experiments, and he discovered another sine qua non prerequisite for success: being present the very moment the young bird pierces the egg. The chick considers the first living being it encounters—bird, human, or other animal—to be its mother or father. Of course, you then have to lavish constant quasi-maternal care on the

▲ Traveling shot from a boat and ▼ from an ultralight, New York.

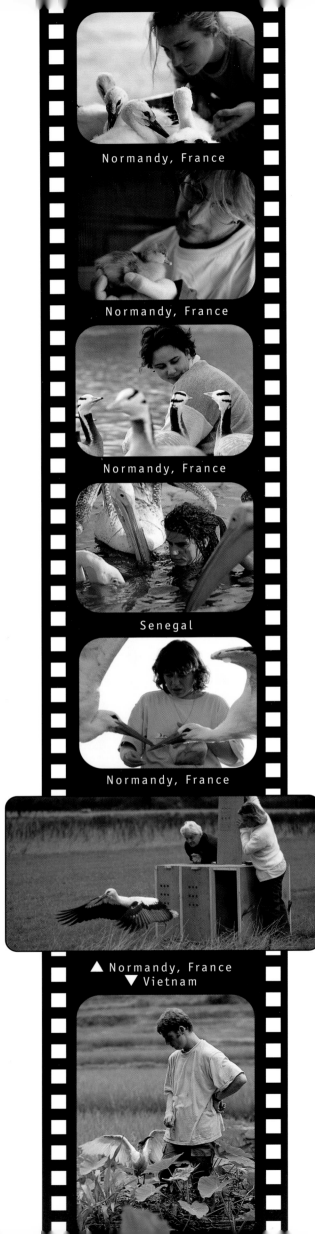

Normandy, France

Normandy, France

Normandy, France

Senegal

Normandy, France

▲ Normandy, France
▼ Vietnam

goslings throughout their youth, sleeping with them, feeding them from your hand or mouth, even swimming with them. As a result, Konrad Lorenz's young geese answered to the voice and gestures of their adoptive father, faithfully following him everywhere on land and in the water—a complete success that had nothing to do with previously known means of training and domesticating birds.

Since this heroic period, these techniques have evolved and have been progressively applied to other species, almost all geese, which quickly grow used to humans since they are precocial. These practices had already been widely used and perfected worldwide by researchers, zookeepers, and even enthusiastic amateurs, but they had not yet been taken to their ultimate logical and natural conclusion. If birds blindly follow their adoptive parents on land and water, then presumably they would do so in the air . . . .

In the 1980s, Canadian Bill Lischman, a pioneer of ultralight flight, was the first to fly with wild geese spontaneously attracted by his miniature aircraft. He decided to raise geese of his own so he could fly with them, a dream that came true in 1988. And it was the short amateur video of him flying with his goose squadron that gave Jacques Perrin, awestruck by this wondrous world premiere, the idea of using ultralights to fly with migratory birds—only ones specially designed for filming them in optimum conditions. It was also Bill Lischman's film that gave

him the idea of flying with other wild species such as cranes, pelicans, storks, wheatears, and macaws, thereby using these birds as "actors" in the film he wanted to make.

So Jacques Perrin contacted a team of aeronautical engineers headed by his namesake, Jacques Perrin at the French engineering company Thompson, and technicians from the movie industry, to investigate the feasibility of this unprecedented project. And, with a team of people who all shared his passion, the producer of The Monkey People and Microcosmos created the first "nursery school–airbase–movie studio for birds." With advice from the above-mentioned French and foreign scientists, but also from environmentalists, veterinarians, and directors of bird parks and sanctuaries, Jacques Perrin created the first avian aviation school in the world from scratch.

The training camp for birds got going in 1998 at Bois-Roger, in the peaceful pastures of Normandy, an idyllic location chosen for its mild, temperate climate and other conditions favorable to the raising of bird species from the four corners of the world. And, perhaps as a lucky omen, this region of lush meadowland also happens to be on a geese-migration route. But Bois-Roger was chosen above all because it offers the optimum conditions for creating pools, ponds, aviaries, and landing strips.

In all, some thirty species of birds, geese, ducks, swans, and even wheatears and swallows

▲ Canoodling with a young greylag goose.

▼ Relaxing with red-breasted goose chicks.

were chosen by the scientists in charge of the project. Meanwhile, a vet working for the National Museum of Natural History, Dr. Eric Plouzeau, set up Bois-Roger's health-management program. At the same time, this specialist in wild birds in captivity also trained the center's staff in hygiene and bird rearing and played a vital role in obtaining the egg-and-chick importation permits necessary to get the program started.

The pride of Bois-Roger, the artificial incubation room, consisted of incubators and ultramodern hatching machines. The futuristic hatchery also included computer equipment specially designed to track and control the chick's birth—the first time this had ever been done.

"To raise birds, we had to have eggs," recalls Marc Crémades, who was in charge of the center and its logistics. "We procured these eggs in France, from wildlife parks such as Villars-les-Dombes, and from all over the world. And, of course, the collecting of those eggs and their importation into France was done with the utmost respect for Customs and Excise, plus sanitary, veterinary, and ecological regulations. What's more, we ourselves imposed our own ethical rules for not disturbing birds, their environment, and, of course, their reproduction. I personally went to several countries—to Iceland, for instance—for the whooper-swan eggs and to Senegal for the pelican eggs."

And only once these priceless and fragile eggs had gotten through the administrative and human obstacle courses placed in their path did they finally reach Bois-Roger. Specially designed "in-flight incubators" were used to transport the eggs at the correct temperature by air, and over land and sea, and, even before it had hatched, each egg already had a thick Customs and veterinary file, full of certificates and official stamps.

"At Bois-Roger, we invented CAI—computer-assisted incubation—and copyrighted the software," says Romain Bianchon, who was responsible for wild swans. "Everything was painstakingly recorded, species by species: the eggs' temperature, hygrometry, and their contents (by candling), data that will provide valuable information for future researchers."

The main qualities required for looking after these birds are a love of animals, patience, calm, a team spirit, and boundless enthusiasm. Another essential ability: to often make do with only a few hours of sleep. Qualifications included no specific diploma, but studies related to the wildlife professions and biology were considered vital. The résumés of several hundred candidates who had heard about the project by word of mouth and through bird-protection organizations, but above all from the unusual job listings posted at the local job center in Caen, reached the offices of Galatée Films in Paris.

In 1998, the first bird nannies took up their posts by their assigned incubators, eagerly awaiting their newborn barnacle, graylag, bar-headed, and red-breasted goose chicks.

▲ Swimming with young whooper swans.

▼ Training common cranes in a Normandy meadow, France.

▲ Display of joy by a whooper swan.

"No one had any experience with these techniques," remembers Romain. "We went about things very empirically, following the instructions of people more qualified than us. We also, of course, learned a great deal from previous experiments, such as those carried out by the National Institute of Agronomic Research. Two days before the hatchings, we looked on, moved, as the chicks tried to break out of their shells. Each one already knew exactly what to do and how. To accustom my baby swans to my voice while they were still in the egg, I read *The Alchemist,* by Paulo Coelho. And, apparently, it worked."

The unusual profession of bird raiser, created by the unique needs of *Winged Migration,* was exercised for three years with dedication, passion, and love by some forty young biologists and adoptive parents. There were two categories: those with responsibility, as a team, for raising and training birds living in a group such as the barnacle geese and cranes—the majority of Bois-Roger's residents—and those looking after an individual bird, such as the mallard.

But since the facilities at Bois-Roger soon proved to be insufficient, two other centers had to be opened abroad, one in the Adirondacks, near New York, for the Canada geese, and another near Phoenix, Arizona, for the snow geese.

Céline Le Barz remembers, "Your heart would literally melt when eight squawking, waddling baby geese gathered round you."

"I spent several nights with the chicks," recalls Aude Mesnil, in charge of the snow geese.

"I couldn't sleep, because they kept snuggled around my head and neck, all forty-five of them, constantly jostling to get the closest to me."

Yannick Clerquin, a globe-trotting biologist more accustomed to ornithological treks on Crozet Island, soon fell in love with "his" pelicans. "The day I arrived at Bois-Roger in December 1999, Aymé was hatching," he says. "It's impossible to see pelicans hatching in the wild. I was deeply moved. The most important thing with pelicans is to have a feel for them. They recognize their names, and even certain words, such as eat."

Nor will Peggy Alexandre ever forget the wonderful feeling she had soon after arriving at Bois-Roger. "With my white coat on, hardly knowing anything about birds," she recalls, "I touched my first little pelican, the size of a coffee cup, and syringe-fed him fish broth." During the shooting, Yannick, Peggy, and Sarka, a young Dutch biologist, went abroad for months at a time, the three of them staying in a remote part of Senegal, in the Djouj bird sanctuary, then in Kenya. "We were so attached to our birds," Yannick remembers, "we would put up with everything: heat, mosquitoes, illness, bad food, isolation—you name it. When the going got tough, we knew we had to hang in there for them. They had become like our children. And pelicans really are like kids: stubborn, affectionate, greedy— but they can be crafty, lazy, and arrogant." The birds became like children for their foster parents, yet young ones still retaining all their natural

BROKEN BAMBOO

IBISES

characteristics and, above all, their independence, flying as they pleased just for the fun of it!

From the egg, they simply had to be accustomed to the noises of ultralight and boat engines, as well as the bicycle horns and other instruments their caretakers used to call them. Staff regularly played tape recordings of these sounds near the eggs in the incubation room.

"The birds had to program these sounds and become accustomed to them before hatching," explains Luc Coutelle, one of *Winged Migration*'s assistant directors. "Chicks don't know how to do anything when they're born. In the wild, their parents teach them everything. We had to get them to walk in a line behind the ornithologists, who were always dressed in yellow so the birds would recognize them and not follow just anybody. After a few days, we took them out for runs in the fields around Bois-Roger. The young birds were then trained to walk and fly along behind a 4x4. Meanwhile, in the nearby lake, they were trained to fly along behind a boat and then to do the same behind an ultralight. You should have seen the ornithologists screaming their heads off at them: 'Come on, that's a girl, come on, you can do it!' Some even had tears in their eyes when a bird flew for the first time."

The intensive training at Bois-Roger, the other centers in the United States, and the various shooting locations in France and abroad gradually produced avian actors accustomed to the presence of film crews and their noisy machines—actors that were always ready to show off, too.

Aurélie Holley, in charge of the barnacle geese, remembers, "Every time the geese landed near us after a successful sequence, they would parade around, sticking out their chests, honking with contentment. Born movie stars!"

Good actors, yes. But, as quite a few of them proved by disappearing during training or shooting, they were not always very disciplined ones. A few harrowing episodes gave rise to extremely unusual "Wanted" messages in the press, at police and fire stations, and on the radio: "If you see birds of this species [followed by a description], please immediately call this number." These odd appeals always received a massive response from the local population, both in France and abroad, especially from schoolchildren, whose teachers often took advantage of the opportunity to instill love and respect for nature in them, some even organizing full-blown nature hunts for the feathered fugitives. And these ads almost always led to the birds' recovery. Like kids caught playing hooky, there they would be a few hours later, hanging their heads in shame in some farmyard or on a church roof. But not always—sometimes they flew the coop for good.

The birds proved they have an enormous capacity to adapt but also dumbfounded everyone by their incredible ability to turn situations to their advantage. "During shooting, we never

▲ Shooting in Southeast Asia.

▼ Training a whooper swan in the Pyrenees, France.

Libya

Libya

Libya

▲ Libya
▼ French navy, North Atlantic

knew whether we were leading the birds, or vice versa," Jean-Michel Rivaud, head pilot at Bois-Roger, explains. "The smartest quickly understood that they had every interest in flying close to the tips of the ultralight's wing because of the favorable air currents it creates, and consequently got a lot less tired than the others. And every time we took off, it was always the same birds that took up these privileged places—until, that is, another one, waiting for his chance, pinched it."

Jean-Michel, an ultralight test pilot with several thousand hours of flying experience and one accustomed to the most extreme shooting situations, had never before experienced anything as moving as flying with birds. "Each kind of bird has its own type of behavior during flight," he says. "In the air, they really consider you're one of them. And we sometimes took ourselves for birds. Sometimes, one of them would come in close and look at you, as though trying to tell you something. One time, a goose really did speak to me during a flight. We were flying over a forest in the Jura range. Filmmaker Thierry Machado was with me and the birds were flying in formation behind us. One of them came up alongside, squawked at us as if to tell us to slow down, then fell back to its place in the formation again. And I realized that, in fact, I was going too fast. Over the radio, Thierry told me he had had exactly the same impression as me, that the goose had definitely spoken to us."

One of the film's co-directors, Dominique Gentil, confirms this sense: "I spent three months in the American deserts filming the migration of the Canada geese. To determine how to approach them and film them, we first had to learn what speed and altitude they flew at."

But what happened to all these birds once the shooting was over? As Jacques Perrin had planned, he donated them to ornithological organizations that had worked in partnership with Galatée Films, such as the Normandy Ornithological Group, or to carefully selected bird reserves, especially those that had provided the eggs and chicks of certain species. As far as possible, the birds' return to life in the wild remained the prime objective of the heads of Galatée Films' center for raising and training birds. An ornithological sanctuary created in the Languedoc-Roussillon region of France was another of the "retirement homes" envisaged for the film's cast after their acting careers were through.

And these actors, filmed for *Winged Migration* by France's most prestigious wildlife filmmakers, stole the limelight. "*Winged Migration* gave me experiences I would never have otherwise had," remembers Laurent Charbonnier, a wildlife filmmaker who has directed a number of acclaimed movies. Like many others who joined Jacques Perrin's team, he did so despite the director's insistence on a completely novel approach to filming birds and close adherence to a scenario. Another

Foldout: Shooting over Normandy, Iceland, and North America.

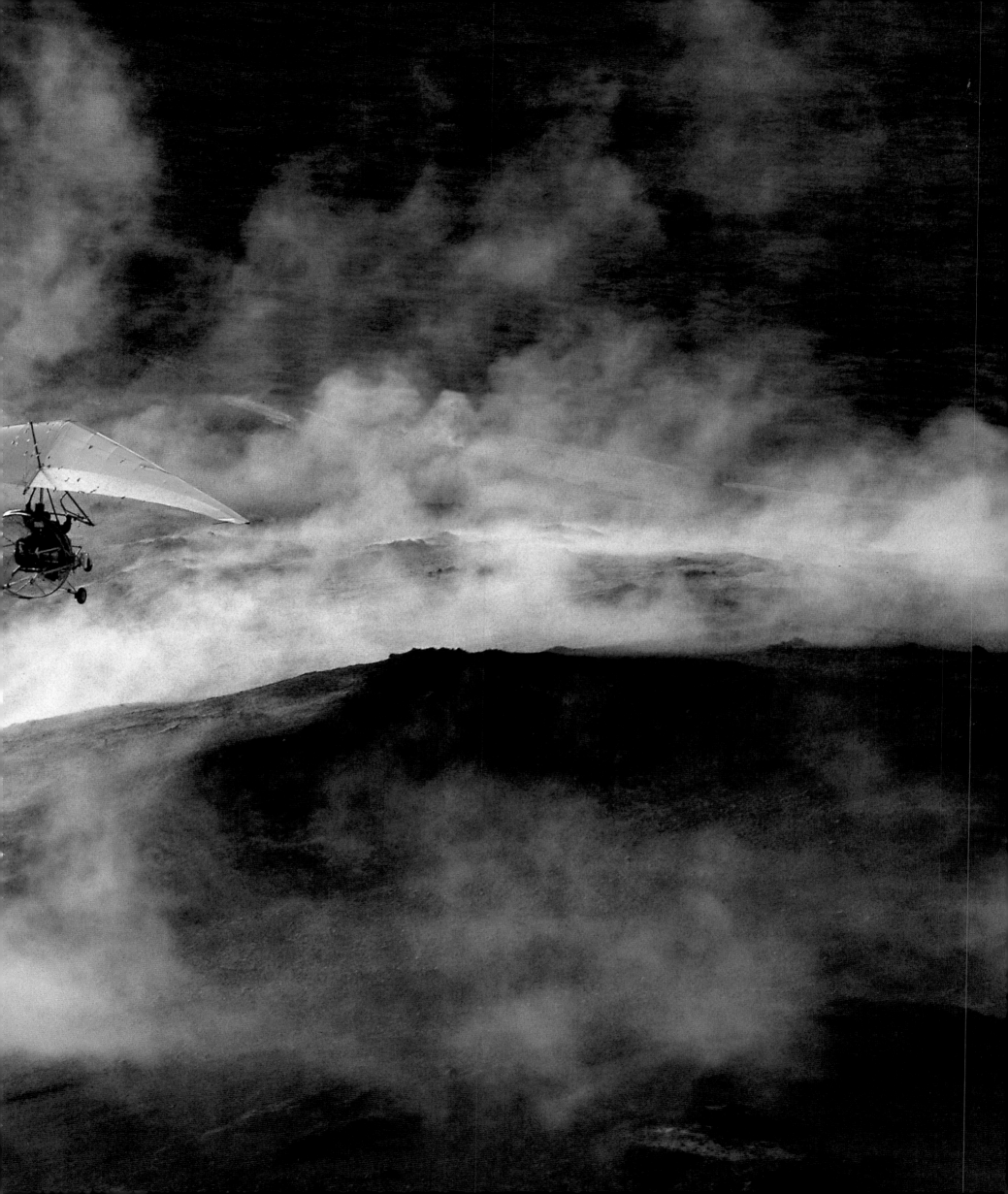

▲ Filming barnacle geese in Iceland.   ▼ Shooting on Lake Powell, Arizona.

SEEN FROM THE FRIGATE – CATCH
THE EFFECT OF THE SWELL. DIFFICULT TRIP
SHOT OF THE SQUADRON.

SHOT OF A FEW OF THEM

EMPHASIZE THE FLYING DIFFICULTIES
(GEESE AND CRESTS OF WAVES)
TURBULENCES CREATED BY THE FRIGATE DOWNWIND

technical constraint for wildlife filmmakers was the decision to use only 35mm movie cameras. These 2½-foot-long heavyweight beasts, weighing between 25 and 50 pounds, plus another 30 pounds of batteries and accessories, are used quasi-exclusively for fiction feature movies and not for wildlife documentaries, whose directors are used to working with lighter, much less cumbersome, and nowadays usually digital equipment.

"The conditions in which *Winged Migration* was filmed, on the frontier between fiction and documentary, imposed very demanding physical and technical constraints," Laurent Charbonnier recalls. "If the script called for mist and snow, then for mist or snow we would sit in our blinds and wait for days—for as long as it took! To be a wildlife filmmaker, you have to like being alone, enjoy silence, and not mind discomfort, and I had my fair share of all that. I shot fabulous, sometimes previously never-filmed scenes on locations almost everywhere in the world. The most moving of all was a huge flight of cranes that came toward me one night in Nebraska. They landed in the mist all around my blind, wings spread, legs stretched out in front. Still moved by the beauty of what I had seen, I told Jacques Perrin on the phone that I had shot the scene in the evening, not in the morning, as the script stipulated. And all he said was, "Well done! So it wasn't in the script, but we'll use it all the same."

To shoot these extraordinary pictures in the air, on land, and on water, the *Winged Migration* crews used the best aeronautical technology available. And often, having tested everything that flies—even the most bizarre machines—they had to invent new ones, given the safety problems and weight restrictions posed by this kind of aerial cinematic acrobatics.

"We envisaged using several types of aircraft—helicopters, paragliders, gliders, paramotors, delta wings, balloons, even remote-controlled mini-gliders or mini-helicopters with remote-controlled cameras," Jean-Michel Rivaud recalls. "Each of these aircraft was suitable for a particular type of bird or movie sequence. A paramotor was used to film hallucinating shots over the cliffs of Skruder off the coast of Iceland. It was also perfect for the unpredictable soaring ascensions of the storks, but it was unusable for fast-flying formations of geese or pelicans. As for balloons, we occasionally used a little airship carrying one camera operator and camera. "But, in the end, it was the ultralight we used the most," he adds. "We had a two-seater prototype made, which we baptized the "ultralight flight pod." Specially designed to carry a pilot and a camera operator, it had a retractable undercarriage to give a 300-degree visual field. For flights over water or snow, floats or skis were fixed to it. In addition, this ultralight had a special engine and an 85-square-foot wing enabling it to fly as slow as 30 mph to film certain slow-flying birds while carrying nearly 1,000 pounds.

▲ Encouraging a Canada goose in the Adirondacks.

▼ Filming a mallard, France.

SUBJECTIVE SHOT: APPROACHING THE AVISO
AND FLYING OVER IT

TRAVELING: THE TWO BARNACLE GEESE WHO TOOK
OFF FLYING ALONGSIDE THE AVISO.

REVERSE SHOT SEEN FROM THE AVISO

No such aircraft had ever been made before. We had to design its structure from scratch. The three of them that were built always performed to our total satisfaction, in locations as varied as rural France, Iceland, and New York. But for bird training, we also used seven other more classical-style ultralights."

The ultralights had to be transported all over the world, and in some countries had to be officially tested to be authorized to fly. And the legal hoops that had to be gone through were sometimes almost as acrobatic as the flights themselves. For example, for a series of sequences shot in America, one of these ultralight pods traveled from Utah to Arizona and then from Arizona to New Mexico before being loaded on a ship bound for Hanoi, with the mountain of previously never-encountered formalities this trek entailed.

On water, too, *Winged Migration* used craft of all shapes and sizes, from the enormous Zodiacs of the Icelandic lifeboat service that transported the film crew and their equipment through raging seas to Skrudur Island to the French Customs and Excise vessels and the navy frigates *Latouche-Tréville, Cassiopée,* and *Loire,* which were only too ready to help Jacques Perrin film the scenes in stormy seas and those of the exhausted greylag geese.

As for shooting on land, the vehicles used for traveling shots often had to be adapted to rough ground on which birds are usually found. A vehicle was even specially designed for the film: a NASA-type amphibious 4x4 robot called the "polypod," capable of taking a camera over soft, marshy, or snow-covered terrain, which worked wonders on the Arguin Banks of the Mauritanian coast. Other less complex conveyances such as a mountain bike often came in handy too, particularly in the Bharatpur Bird Sanctuary in India, where motor vehicles are forbidden.

Innovation was also the name of the game when it came to movie equipment. This included the Wescam stabilized-camera system, a kind of large bubble with a gyroscope inside, suspended beneath a helicopter or ultralight. *Winged Migration*'s pioneering production team devised and built various ingenious systems, ranging from employing a retractable camera boom to using simple luggage elastics to lighten the camera on the ultralight, for which camera operator Sylvie Carcedo became a specialist.

"When confronted with novel technical problems, you have to reinvent everything," explains Alexandre Bügel, one of Galatée Films's head grips. "For *Microcosmos,* engineer Romano Prada had designed an extremely sophisticated robot for shooting tiny traveling shots and filming all kinds of insects. For *Winged Migration,* it was impossible to systematically use a single piece of equipment, since the shooting con-ditions included camera movements in the air, on water, in the sun, in tropical countries, and in the rain, sand, or ice, as well as with species as different as the robin and the

▲ ▼ Shooting on board the French navy vessel *La Loire*.

Iceland

Utah

Normandy, France

Crozet Island

Normandy, France

▲ Normandy, France ▼

albatross. And, as there was no question of resorting to special effects and computer-generated imagery, tools specific to each species and environment had to be designed.

"We made a totally new kind of camera vehicle out of a modified Land Rover, an inflatable camera boat with the camera mounted inside a blind, and even a lightweight camera crane that can be carried on someone's back, plus a special seat enabling one to fix a camera operator on any kind of land vehicle or watercraft," Alexandre Bügel adds. "With Jean-Marc Mouligné, Galatée Films's head technician, we even made a remote-controlled camera mount for the crane and for use in blinds. Unfortunately, for safety reasons, one or two of these technical innovations weren't passed as airworthy.

As the filming went on, we had to improve some equipment, for instance the camera suspension for the camera vehicle, to enable us, for the very first time, to begin with a traveling shot preceding the flying birds and then do a 180-degree pan to one following the same birds from behind. To film albatross in New Zealand, I dreamed up a machine capable of obtaining steady pictures from a boat in high seas, a kind of giant universal joint inside which the camera and camera operator would be suspended—rather like an astronaut's test seat. To film the storks spiraling up on thermals, new fittings for attaching a camera to a paramotor had to be devised. We also designed a gyroscopic camera

mount, a piece of movie equipment still considered extremely advanced."

"We managed to film through every kind of torment, even in the eye of storms," Jacques Perrin adds. "And nowhere does this technical prowess show in the film." But for the director of *Winged Migration,* this feat in no way effaces the accomplishments of all those who came with him on his journey, each of whom played a vital role in helping build that dream with him. "One only has talent through that of others," he emphasizes.

As the months passed, the budget of this extraordinary film went through the roof, until it became one of the most expensive European films ever made. But its producer and director remained serene throughout. Jean de Trégomain, *Winged Migration*'s executive producer, recalls, "When we told him that a shooting schedule was running way behind time because of foul weather, that overheads had quadrupled, he never asked himself what we could do to remain on budget, only where we could find more money."

"Such is Jacques Perrin, one of the last great princes and great dreamers in cinema," adds Michel Debats, who had himself directed *Himalaya* in 1999, before working as assistant director on another saga of the sky.

"One day, I spoke to a young man who had just spent four months on the other side of the world," Jacques Perrin says. "Four months for

▲ Readying to shoot in Libya.

▼ With the pelicans on a beach in Senegal.

two and a half minutes of film—a long, long wait, until one day, in an ink-dark sky over a raging sea, there was a slanting ray of light. The footage was hallucinating. It was as if I was watching an albatross display for the first time. And it was all that time spent waiting for it that had captured that miracle on film."

A single figure says everything about a quest for quality rarely matched in the history of cinema: less than 1 percent of all the footage shot was used in the film—only 0.5 percent, in fact! The ratio for wildlife filmmakers, extremely demanding in this respect, is between 2 percent and 4 percent. For fiction films, it is between 10 percent and 20 percent. But the other 99.5 percent that ended up on the cutting-room floor is priceless. For it represents the unfathomable amount of energy, passion, and talent that goes into shooting the film, as well as the emotions experienced by all the teams of *Winged Migration* in the four corners of the world over a period of several years.

It is this alchemy between land and sky that constitutes the magic of *Winged Migration*.

THE PROMISE OF RETURN

by Jacques Perrin

Launguedoc-Roussillon, Aveyron, Lower Normandy, and Calvados are some of the regions and departments of France on the flight paths of migratory birds and the sites of annual rendezvous for their ephemeral stopovers. Messengers of time regained, celestial guides bringing us tidings of the earthly paradises scattered along their migration routes, they reach us from distant horizons, eternally returning, appearing and disappearing with the seasons, coming and going with the delicate tints of spring and the fiery hues of autumn.

To reach their Arctic homeland or their wintering grounds around the Mediterranean rim, they blaze trails for us through the sky and, with each seasonal return, reveal to us again the timeless beauty of our own familiar surroundings.

The General Councils of Aveyron and Calvados and the Regional Councils of Languedoc-Roussillon and Lower Normandy thought it only fitting that they should assist us in organizing the shooting in their regions. It was their support, and the additional aid of EDF, Lufthansa, and Crédit Agricole, that gave Winged Migration the wings to get off the ground. I would like to express our gratitude to each of these companies.

▲ With common cranes in Languedoc-Roussillon, France.

▲ With barnacle geese over Calvados, France.

▲ With barnacle geese over Mont-Saint-Michel, Lower Normandy, France.

▲ Shooting at Bel Castel, Aveyron, France.

"AROUND THE WORLD IN A WEEK"

by Marc Crémades, head of Galatée Films's bird-raising centers

In the course of my job as head of Galatée Films's various bird-raising centers, it was often my task to transport eggs and birds all over the world. One of my most hectic schedules took me from France to Guyana, from Guyana back to France, from France to Senegal, from Senegal to Germany, from Germany to France, and then from Paris to Switzerland. The only thing out of the ordinary about this itinerary was that, due to shooting commitments, the whole lot had to be done in a week.

It all began with the transportation of our nine hand-raised macaws, which had to be taken to be filmed in a forest in Guyana. Any transportation of live specimens of a protected species generates massive administrative and veterinary formalities destined to prevent wild-animal trafficking—a single hyacinth macaw can fetch traffickers between 7,500 and 15,000 dollars—we had to go through all the necessary procedures to export these protected birds.

Macaws are very sensitive to temperature changes. After checking with the crew that the temperature in the baggage hold was indeed 68° Fahrenheit, we took our seats on a night flight to Guyana. On arrival, we had to help the crew already on location finish the aviaries before taking the night flight back to France. When I arrived at Roissy, I managed to get on another night flight to Dakar, where I had to join our pelican crew and its twenty charges, which had just finished shooting in northern Senegal. In Dakar, we went through the delicate export procedures with the Customs forwarding agent and the Senegalese veterinary authorities before the birds' departure.

The pelicans arrived dehydrated after a long trip by truck and required lengthy attention. That evening, we loaded them onto a Lufthansa freight plane that arrived at midnight in Frankfurt, where again the birds had to be hydrated and fed. At 6 a.m., seven of them, not operational for the scheduled shooting, returned to France, where they were taken to the National Museum of Natural History in Paris by one of our bird teams. The remaining pelicans and I then got on an overnight freight flight to Nairobi, accompanied by Yannick Clerquin, the biologist who had looked after them from birth. Throughout the flight, with the captain's authorization, we went into the plane's freight area to check the state of the birds—on one condition, that we wore oxygen masks just in case.

When we arrived in Kenya, once we had attended to another mountain of official paperwork, the pelicans were taken in three trucks to Lake Magadi, a two-hour road trip south from Nairobi, where a big problem awaited us: the birds refused to eat. They would not touch the fish that had been provided for them by the local crew. So we had to go looking for tilapias, standard pelican fodder in Senegal, and, finally, after a two-day search, we stumbled on a supply. A restaurant owner put us in touch with his supplier, a local fish farmer, who was flabbergasted that we wanted to buy only his smallest fish, and such a huge quantity. Meanwhile, due to this food "hiccup," our shoot had been severely jeopardized and, above all, the life of our beloved pelicans endangered.

Minutes after we had arrived back in Nairobi, I was on a plane bound for Paris. And as soon as I arrived at Charles de Gaulle International Airport, I left for the Alps, where we were filming bar-headed geese at 11,500 feet.

Helicoptering equipment on Bylot Island in the Canadian Arctic.

▲ On a beach in Brittany, France.

▼ Approach to a shooting location in Nepal.

"WHEN BIRDS FLY IN FORMATION, THEY ECONOMIZE ENERGY"

by Henri Weimerskirch, CNRS Research Director at the Center for Biological Research, Chizé

The shooting of Winged Migration had unexpected scientific spin-offs, one of which was that my team became the first to study the energy expenditure of pelicans during flight.

Why do many species of large birds fly in V formation or in rank? There are several hypotheses: flying like this enables them to reduce the energy cost of their flight, it allows them to remain in contact with one another, or it helps improve their navigation by communicating.

The first hypothesis is the one most scientists favor. There have been numerous theoretical aerodynamics experiments but, curiously, no field study has been carried out to directly verify it. Theoretically, all birds flying in formation except the leading bird benefit from the turbulences produced by their neighbors' wings, and therefore use less energy—just as the fuel consumption of a plane flying at the rear of a formation is 20 percent lower. And, theoretically, for maximum reduction of energy consumption, birds flying in formation would have to fly at a very precise distance from and position vis-à-vis the one preceding it.

Observation of wild-bird formations in nature, however, very frequently shows that they do not fly at the distances and in the positions predicted by theoretical models. Not only that, but they often fly in positions where their energy expenditure is increased. The energy-consumption hypothesis, therefore, still had to be proved, and the only means of doing so was by comparing, as with planes, the energy consumption of birds flying in formation with those flying alone. And this posed as much a technical challenge as a logistical one: How, in the wild, can one measure the "fuel consumption" of a flying bird? Yet this is exactly what the filming of Winged Migration enabled our "pelican project" to do.

First, though, how exactly does one measure a bird's energy consumption?

The "cost" of flight is generally measured in relation to a bird's energy metabolism at rest; the standard of measurement is oxygen consumption and carbon-dioxide production. When an animal produces an effort, its breathing and pulse rate increase. A bird, therefore, uses up between two and ten times more energy flying than it does at rest. But there are a great many different kinds of flight, ranging from wing beating, at a rate of seventy times a second in hummingbirds, to the gliding flight of large birds of prey.

There had previously been only one laboratory technique to measure the energy expenditure of an animal: measuring its gas exchanges (the consumption of oxygen and the rejection of carbon dioxide). This was done in special metabolic chambers whose gas intakes and outputs can be directly controlled, or sometimes by using masks linked to analyzers (a technique that enabled the first-ever measurement of the energy expenditure of a bird in flight in a wind tunnel).

But these were merely laboratory experiments. How could one measure the same energy expenditure in the wild? Recent advances in microelectronics have made this study possible. By measuring the pulse rate of an animal in controlled laboratory conditions, it was shown that pulse rate is directly linked to oxygen consumption. So we decided to adapt the recorders used by athletes—the unit and its electrodes weigh under 2 ounces—for use on large birds such as the albatross, and we were able to show that large albatrosses use up very little energy flying. When they fly with favorable winds, they use up only 10 percent to 20 percent more energy than they do sitting on the ocean. Taking off, on the other hand, during which a bird has to run on the water into the wind for perhaps 30 yards, uses up a great deal of energy.

While this technology was being tried out on albatrosses on Crozet Island, I had the chance to view the first rushes of Winged Migration, superb footage of geese in Iceland shot from an ultralight. And it worked: from an ultralight one could observe in detail geese flying in formation and measure their respiratory rhythm and wing-beat rate. All one had to do to study their energy expenditure in different flight conditions was to equip these tame birds with a pulse-rate monitor.

Unfortunately, the albatross study was never carried out, as the reduced size of the equipment would not have enabled its optimum performance. So pelicans, the same size as albatrosses—15 to 22 pounds—were chosen instead.

The study was carried out in the Djouj bird sanctuary in Senegal, where, for the scheduled

▲ Traveling shot in Senegal.

Pelicans in Senegal.

Ready for takeoff.

shoot, one of the Galatée Films teams was training pelicans to fly in formation behind a boat along the river or in the air behind an ultralight.

The first thing we did was measure each pelican's pulse rate at rest, which is 80 beats a minute. Our first surprise was to find that the pulse rate, and therefore the energy consumption, was higher when a bird was walking or swimming than when it was flying. These two types of activity engage different muscles: highly developed pectorals for flight and leg muscles for walking or swimming.

We also discovered that during solo flights behind the ultralight, outside its slipstream, of course, a bird's wing-beat rate is higher than 90 per minute, whereas in formation it beats its wings only an average of 62 times per minute. And, during formation flights, each birds beats its wings either in unison with the group leader or after the bird in front of it, each wave of wing beats beginning with the leader.

Live pulse-rate measurements made using miniaturized electronic recording devices also showed that a bird flying in formation economizes between 15 percent and 25 percent of its energy compared to solo flight, and 25 percent to 30 percent when gliding. These real-life energy savings therefore confirmed the results arrived at using theoretical models, and formally established that pelicans economize energy by flying in formation on their long migratory flights, during which they often cannot feed, especially when crossing desert regions. These results clearly need to be refined and applied to other species, but thanks to the birds of Winged Migration, the in-flight study of flight energetics is now a reality.

"WITH THE BIRDS, I FLEW OVER THE MOON"

by Jean-Michel Rivaud, *Winged Migration*'s head pilot

Winged Migration *enabled me to realize the dream of many a pilot: to fly
over the most beautiful landscapes in
the world in an ultralight, yes, but with birds.
And to share three years of exceptional experiences and so many rich moments in only a few
lines is all but impossible.*

*One morning, after ten days of foul weather,
we were flying across the side of a volcano in
Iceland, a tongue of black, granular, blistered
lava from which wisps of smoke were occasionally
rising—a disturbing lunar landscape. I had been
flying with assistant director and camera operator Thierry Machado for twenty minutes, between
3 and 10 feet above the ground with the undercarriage retracted, at the speed of the geese we
were going to film. We were waiting for the right
light, the temperature was about 40° Fahrenheit,
and there was the constant anxiety of engine
failure. In conditions like these, not only does a*

*flight seem to drag on, but you also begin to
wonder just what you are doing in such inhospitable country.*

*And then, suddenly, the sun appeared just
above the summit, bathing the birds in light,
illuminating only the wisps of vapor and a few
hummocks. Magically, hell had suddenly become
heaven, and Thierry's skill had done the rest.
"It's in the can!" The interlude had lasted at the
most three or four seconds. Powerless to express
the emotions we felt, we led the birds back in
silence.*

*Again in Iceland, we lived to tell the tale
of another extraordinary experience, on the
Vatnajokull glacier, which descends 7,000 feet
directly into the Atlantic, where its seracs tumble
into the sea to become icebergs. The idea was to
discover with the bar-headed geese the wondrous
beauty they fly over during their migrations.
For once, flight conditions were perfect, the light*

*and aerology were ideal, and there wasn't the
slightest turbulence. And then, suddenly, contrary
to habit, the geese did a U-turn and flew back
toward the team on the ground before we could
stop them. Moments later an extremely rare and
totally unpredictable phenomenon was upon us:
an enormous pocket of cold, dry air came down
the glacier, mixing with the warm, humid air
coming off the ocean, and, for a few minutes,
helpless, not understanding what had hit us,
we were buffeted about by terrifying gusts of
wind and turbulences of unimaginable violence.
Suddenly everything was calm again. Having
never encountered anything like this before,
there was no way I could have anticipated it.
Though the geese, with their mysterious, sixth-
sense sensors, had been able to.*

"THREE YEARS OF THRILLS"

by Laurent Charbonnier, head wildlife camera operator

When Jacques Perrin asked me to shoot certain sequences of Winged Migration, I had no idea that for the next three years, I would live through some of the most wonderful experiences in my life as a wildlife cameraman. Here are a few flashbacks:

• **February 1999: Most glacial blind**

It was in Japan. We were shooting the wild swans. It was so cold, the camera wouldn't roll.

• **March: Biggest disappointment**

Waiting at Kushiro Airport on Hokkaido Island for a plane back to France via Tokyo, we sat watching the snow falling outside in great big flakes—the snow we had waited three weeks for to film the dance of the Japanese cranes, and which had come at last.

• **March again: Longest stakeout in a blind**

I had fourteen airline tickets in my pocket when I left for the United States—and I can't stand flying! In Nebraska, I had to film a Canadian crane "dormitory." They're very fierce birds. We dug a blind in the dirt, covered with planks and sand, on an island in the Platte River, where, every evening, tens of thousands of cranes came to sleep. The first began to arrive around 5 p.m. Having spent the night in the blind, at daybreak we discovered an extraordinary spectacle: the cranes were everywhere around us, some only 15 feet away!

• **Late March: Most magical shoot**

Farther north, in the Rockies, we had been shooting the sage grouse sequence for six days, but to no avail: foul weather, a resident buzzard who scared off the grouse, and so on. And then, one morning,

a miracle happened: magical light, no buzzard, and we shot the sequence in less than an hour.

• **April: The catapult blind**

This time, we were in southern Oregon to film western grebes, Ross's geese, and pelicans. Pelicans are very fierce, so we decided to use a catapult to attract them. From the blind, we catapulted small fish and pieces of bread, and ended up attracting the seagulls, whose hullabaloo attracted the pelicans.

• **May, June, and July: 4 inches from the cuckoo**

We were in Sologne, in a marsh belonging to the Château de Chambord. I finally found a warbler's nest with two eggs inside, and, as I was hoping, a cuckoo's egg was soon laid in it. When the warbler had laid all her eggs and begun sitting on them and the cuckoo egg, I began sitting in my blind. The cuckoo chick very soon hatched, and for twenty-four hours nothing happened. The next day, at daybreak, the baby cuckoo began busily heaving the other eggs out of the nest. But—calamity—my camera wouldn't work! The blind had gotten too damp the day before! There was no way I was going to miss the incredibly rare opportunity to film a baby cuckoo at work. So I hopped in the car, drove to Paris, had the electronic components of my camera changed in five minutes, and was back at Chambord in a flash—just in time to film the baby cuckoo still moving eggs!

• **August, September: The shoot we should never have done**

I was filming migrating waders in the Wadden Sea, where we had spent twenty-one fifteen-hour days without shooting any decent footage—three weeks

▲ Shooting on the edge of a marsh, India.

▼ On the lookout for bald eagles, Alaska.

wading about in mud and slime, up to our waists in seawater, for nothing, or almost nothing.

• November, December: The shoot when we drank only water

We were in Mauritania to shoot millions of waders, based between the desert and the sea—i.e., in the middle of nowhere. Thanks to a bizarre piece of equipment specially designed for Galatée Films, we were able to shoot wide-angle shots of birds from only a few feet away. To get around, we used the sailboats the Imraguen use for fishing. And, to further complicate matters, the birds were constantly moving about with the tides, and, more often than not, we had to transport the equipment on foot, tramping for miles through mud.

• January: The shoot that should have been easy.

In the Camargue, we spent whole days in freezing water watching wary geese who were never in the right place.

• February: Noisiest shoot

Back in Normandy. Five hundred thousand starlings arriving to sleep in a marsh flooded with very cold water is an extraordinary spectacle. And, even more impressive, in the morning, we witnessed mass arrivals and departures that lasted only minutes.

• February, March: Hottest shoot

The place: Lake Bogoria in Kenya, temporary home for a million flamingos, many of whom died from the pollution and were devoured before our eyes by eagles, baboons, and marabous.

• April, May: Simplest shoot

Without a break, there we were in Quebec, where millions of snow geese stop to recharge their bat-teries on the banks of the Saint Lawrence. We shot several shots from a blind. There were long waits, but the geese arrived as expected.

• July, August: Windiest shoot

To get to the very difficult-to-get-to small island of Skrudur off the coast of Iceland, we did the 3-mile crossing in a Zodiac. About a million birds nest there. We had to rope down to film the guillemots, puffins, and gannets. The wind was so violent, we could hardly stand up on the cliff edge.

• August, September: Shortest shoot

We were in Sologne for the departure of the swallows, and, in the bat of an eyelid, off they had gone, already too far away.

• October, November: Most cinematic shoot

Hunting for woodpigeons with nets in Spain.

• December: Most vertiginous shoot

In the Grand Canyon, in a cold sweat, filming bald eagles.

• January: Most pleasant shoot

Our 5:30 a.m. flight to India on January 1 put a damper on New Year's Eve. Destination: Bharatpur National Park, south of Delhi. There, in a misty marsh, we had to film the displays of antigone cranes, eagles, pelicans, egrets, ibises, and spoonbills. Cars were strictly forbidden in the park, so, for twenty-three days, we went everywhere on bicycles.

• February: The shoot when I wasn't supposed to shoot birds

In the forest of Chambord, we again dug a big hole, installed a false lens sticking out of the top of the blind, and put out corn each day to attract stags.

Three weeks later there they were, eating less than 3 yards from the camera.

• February, March 2001: Back in Sologne

Mallards mating, coots fighting. Life goes on, and so does filming.

At this moment, I still had a few more months to spend in the wonderful wake of Winged Migration.

Crow's-nest camera position for macaws in Peru.

▲ In the midst of whooper swans on Hokkaido Island, Japan.

Birds tomorrow

"Watch them fly past, the wild ones
Going where whim takes them
Over mountains, woods, seas, and winds
Far from all slavery
The air they drink would burst your lungs."

Jean Richepin, *Birds of Passage*

Indian sunset.
———
Green-winged macaws.

From the dream of Icarus to that of Leonardo da Vinci, humans have always been fascinated by birds. For centuries, we have strived to imitate them, to join them up into the air—in vain.

Birds frequently figure in the most ancient mythologies, either as quasi-gods or as gods such as the ancient Egyptian falcon god Horus, the Roman goddess of wisdom Minerva, the Aztec god Quetzalcoatl, and the Celtic crow divinity. Herodotus describes how the red-and-gold plumaged phoenix was consumed by fire, then rose again every 500 years.

A re birds a symbol of eternity?

They have been worshipped for millennia, and still are in parts of Africa, Oceania, the Americas, and Asia—as if these feathered, discreet, and regally free creatures hold secrets older than humanity itself. Many a so-called primitive society has had ceremonies celebrating birds. For example the Easter Island cult of the birdman endured until the late nineteenth century. Birds, venerated since the dawn of humanity, seem always to have served as intermediaries between humans and their gods, between the living and the dead. In the Old and New Testaments, winged angels and archangels such as Michael, Raphael, and Gabriel—half birds, half men—were charged with bringing God's tidings.

Until modern times, the vital equilibrium between humans and birds was more or less respected. It was sometimes, however, severely compromised, due to economics (the abusive killing of birds for food) or politics (Chairman Mao's decree that peasants eliminate every sparrow to protect China's harvests).

Despite these past aberrations and the disastrous overhunting that continues today, up until only a few decades ago humanity's relationship to the world's avifauna remained one of respect and even concern.

What about today? Population explosion and the headlong industrialization and urbanization of the planet has inevitably brought pollution and the yearly diminution of wildernesses and therefore the territory of birds. As *Winged Migration* unflinchingly shows, birds are today endangered as they have never been previously in the millions of years they have lived on Earth.

Our hope is that Jacques Perrin's film and this book can help us better understand birds, better love them, and better protect them. Birds may not be the messengers of the gods any longer, but they will always remain harbingers of happiness. For they come from afar, braving every peril, to bring us their fascinating *joie de survivre*.

Two king penguins en route for the high seas, Falkland Islands.

Olli Barbé

Dominique Gentil and Jacques Perrin

Thierry Machado

Luc Drion

Laurent Charbonnier

Philippe Garguil

▲ Laurent Fleutot
▼ Sylvie Carcédo

Credits

A FILM BY JACQUES PERRIN
Codirected by Jacques Cluzaud and Michel Debats.
Cowriters of the script outline: Stéphane Durand
in collaboration with Jean Dorst, Guy Jarry, and
Francis Roux (scientific advisers).
Assistant directors and cinematographers: Olli Barbé,
Michel Benjamin, Sylvie Carcedo, Laurent Charbonnier,
Luc Drion, Laurent Fleutot, Philippe Garguil,
Dominique Gentil, Bernard Lutic, Thierry Machado,
Stéphane Martin, Fabrice Moindrot, Ernst Sasse,
Michel Terrasse, Thierry Thomas.

EDITING/POSTPRODUCTION
Film editor: Marie-Josèphe Yoyotte.
Assisted by Colette Beltran, Pauline Casalis, Séverin,
and Catherine Mauchain.
Sound editor: Gina Pignier. Assisted by Michel
Crivellaro and Michel Trouillard.
Mixing: Gérard Lamps. Assisted by Armelle Mahé.
Sound effects: Laurent Levy.
Special-effects coordinator: Alain Le Roy.
Laurent Quaglio.

MUSIC
Bruno Coulais.

PRODUCTION
Producer: Jacques Perrin.
Acting producer: Christophe Barratier.
Executive producer: Jean de Trégomain.
Public relations and partnerships: Yvette Mallet.
Press attaché: Eva Simonet.

PRODUCTION MANAGEMENT
Philippe Gautier.
Philippe Baisadouli, Claude Canaple, Jean-Michel
Deroche, Christian Drozdzik, Catherine Pierrat,
Stéphane Quatrehomme, Vincent Steiger, Charles
Stenhouse, Xiaoling Zhu-Pradinas.
Jean-Philippe Avenel, Severin Bouyer, Luc Coutelle,
Gil Descoing, Pierre Ferrari, Paula Luttringer, Boucif
Mohammedi, Johann Mousseau, Sandrine Morvan,
Thomas Pflimlin, Vincent Piant, Hélène Rio, Franck
Vadé, Pierre Vaysse.
Vincent Allard, Emilie Allera, René Autier, Stéphane
Barlow, Isabelle Baylli, Olivier Boucknooghe,
Laurence Cénèdèse, Corentin and Pierre-Emmanuel
Chaillon, Frédéric Chartier, Marie Dumont, Xavier
Fabre, David Forax, Jean, Jacky, and Philippe
Gardette, Frédéric Grousseau, Frédéric Hornecker,
Loïc Jouanjan, David Juhere, François Le Flamanc,
Pascal Legal, Bruno Léger, David Lemenan, Gérard
Lengaigne, Philippe Louis Lalande, Lotta Nilsson,
Damien Lutringer, Olivier Meunier-Colin, Karl

Meriais, Pascal Mousselet, Philippe Nicolas,
Christophe Quirin, Cedrik Rabier, Gilberto Rodrigues,
Christophe Siari, Bernard Simonnet, Sophie Stenhouse,
Fabrice Triquenot, Cédric Vignand.
India: Atul Srivastav; **Bylot:** Odile Dumais; **Canada:**
Brian Black; **Crozet:** Peter Crunelle, Frédéric Essob;
Kenya: Ross Withey; **Libya:** Noureddine
Aboukarwata; **Nepal:** Pasang Kipa Lama, Champa
Kalsang Tamang; **New Zealand:** Olivier Rochery;
Senegal: Lamine Fall; **United States:** Robin Slosser,
Robert Solberg, Sisco Orville; **Vietnam:** Pham Van
Ba, Tran Hoa.

ADMINISTRATION
Administrators: Chantal Cohen-Touboul, Paulette
Materne, Claude Morice. Assisted by Auriane
Bonalair, Fabrice Corniglion, and Patricia Maternik.
Production secretary: Patricia Lignières.
Assistants: Nasser Belkalem, Nicole Devaux, Claire
Dornois, Magali Herbinger, Nicolas Mauvernay,
Zahia Moudres, Jean-Luc Tesson, Karine Tourgeman.

DIRECTOR'S TEAM
Assistants: Jérémie Appery, Élodie Baticle,
Emmanuelle Debats, Marie Miquel, Ursula Sigon,
Hadrien Soulez Larivière, Frédéric Vignal.
Second-unit director: Zhang Xian Min.
Ornithologist and assistant: Guillaume Poyet.
Storyboard: Olivier Chérès.
Nautical adviser: Bernard Deguy.

SCIENTIFIC CONSULTANTS
Kenneth P. Able (United States); Peter Berthold
(Germany); Hiroyoshi Higushi (Japan); Yossi Leshem
(Israel).

CINEMATOGRAPHY
Camera operators: Éric Catelan, Alain Ducousset,
Benoît Nicoulin, Lee Parker, Zhang Yuan.
Assistants: Christophe Adda, Russ Allinson, Michel
Arabeyre, Marc Asmode, Jean-Paul Agostini, David
Aïm, Pierre Bec, Pierre Berthier, Sarah Bouyain,
Guillaume Brault, Catherine Briault, Claire Caroff,
Pierre Chevrin, Isabelle Czajka, Jean-Yves Delbreuve,
Dominique Delguste, Jean-Marie Delorme, Nasr
Djepa, Isabelle Dumas, Stéphanie Etevenon, Pascale
Ferradini, Nicolas Gaurin, Tiphaine Helary, Maxime
Heraud, Colin Houben, Xavier Jalain, Maxime Jouy,
Alexis Kavyrchine, Michel Koelblen, Vincent
Kotwas, Sylvain Maillard, François Mestoudjian,
Christine Mignard, Christophe Pérotin, Sylvie Petit,
Christophe Pottier, François Quillard, Éric Sicot,
Rodolphe Soucaret, Pierre Stadnicki, Roger-Paul
Tizio, Thierry Tronchet, Thierry Taïeb.

Marion Befve, Éric Bonnot, Guillaume Brault, Thomas Bresard, Maryse Charbonnier, Tony Chapuis, Pierre Chevrin, Nicolas Cornut, Sarah Couvelaire, Thibaud Danton, Nicolas Deblonde, Élodie Delettre, Hervé Kern, Sophie Leonetti, Cyril Lèbre, Mathilde Louveau, Alain Lutic, Stéphane Patti, Frédérique Saj, Pauline Teran, Anna-Katia Vincent, Lorenzo Weiss. Head of photography: Mathieu Simonet. Behind-the-scenes footage photographers and directors: Patrick Chauvel, Renaud Dengreville, Toinette Laquière, Hugo Levy, Florent Marcie, Renan Marzin.

SOUND
Head sound engineer: Philippe Barbeau. Jean-Baptiste Benoît, Denis Guilhem, Paulo De Jesus, Philippe Lecocq, Marc Soupa, Martine Todisco, Kathy Turco. Assisted by François Bidart, Camille Chenal.

MACHINERY
Sylvain Bardoux, Vincent Blasco, Olivier Bouysson, Alexander Bugel, Gilles Cousteix, Jean-Yves Freess, Roland Gautherin, Florent Geslin, Thierry Pascal, François Perrault-Alix, Philippe and Roger Priot, Étienne Saldes. Assisted by Rodolphe Albarede, Jacques Bruzeaux, Éric Fodera, Nicolas Gayraud, Pascal Ghristi, Sergeï Lourié, Julien Rohel, Jacques Baloffi. Technical constructions: Jean-Marc Mouligné and Xavier Morin.

ELECTRICITY
Olivier Barré, Richard Brodet, Guillaume Brunet, Gérard Caumon, Martial Combe, Michel Foropon, Virgil Gomez, Olivier Guarguir, Michel Lefrançois, Nicolas Maigret, Denis Moncel, Younès Najfar, Philippe Porte, Patrick Vobecourt. Assisted by Gregory Alonso, Damien Combe, Pascal Cormont, Mathieu Gheux, Cafer Ilhan.

SET DESIGN
Régis Nicolino. Assisted by Bernard Ducrocq, Mikael Deschamps, Jacques Taillard. Xavier Buffin, Dominique Helias, Patrick Gaslène, Jean-luc Roselier, Jean-Paul Tanchereau, Valérie Amoussa, David Gonçalvès, Julien Harant, Robert Mougin, Pierre Jouille. Assisted by Daphnis Lenoir, Jean-Pierre Maricourt, Philippe Tessier, Christian Renaud, Guy Vanderplaetsen.

PILOTS
Head pilot: Jean-Michel Rivaud. Ultralight pilots: Marc Bruckert, Hervé Cousquer, Philippe De Cressac, Frédéric Cruciani, Christine Desdoit, Gilles Desheulles, Jean-Patrick Deya, Luc Dulude, Bernard Dupont, Alain Feuillette, Scott Johnson, Jean-Luc Laine, Serge Mesnard, Laurent Patte, Cédric Poyet, Edgar Raclot, André Saint-Germes. Balloon pilot: Alain Aubry. Paramotor pilots: Alain Arnoux, Philippe Dessaigne, Mathieu Dottori, François Lagarde, Thierry Mazzarelli, Michel Touitou. Wescam pilot: Dany Cleyet-Marrel. Boat helmsmen: Éric Chevalier, Olivier Kerael.

BIRD-RAISING AND BIRD-TRAINING CENTER
Head: Marc Crémades. Peggy Alexandre, Karine Ancrenaz, Isabelle Ange, Romain Bianchin, Julie Bernouis, Grégory Besnard, Yves Bion, Mickaël Bordeaux, Frederik Burke, Mickaël Camus, Jocelyne Caumartin, Emmanuel Cavelier, Hélène and Bertrand Chauvel, Yannick Clerquin, Yvin Dagorne, Dann, Guillaume Delaunay, Evangéline Depas, Alassane, Baye, Racine Diop, Ahmed Diouf, Marie-Noelle Divet, Cyril Drieu, Irakli Ebralidze, Aurélien Gallier, Christopher Goodfellow, Caroline Groussain, Julien Harant, Frédéric Hare, Rachel Henriques, Vanessa Hequet, Hoa, Aurélie Holley, Julien Houeix, David Hubert, Clémence James, Clémence Jarry, Sarka Jiraskova, Joseph, Antoine Journe, Antoine Julien, Céline Lebarz, Charlotte Leman, Patrick Lelievre, Bénédicte Lericolais, Jennifer Liegeois, Karine Limanton, Tony Mauger, Stéphane Mautref, Fabien Menanteau, Aude Mesnil, Caroline Molliet, Georges Ovaschvili, Georges Pareshishvili, Lisa Pecullo, Nicolas Petitou, Frédéric and Christophe Profichet, Quan, Jérôme Raynaud, Didier Reynard, Sophie Royer, Christelle Signol, Sonko, Richard Stenhouse, Tuan Tong, Myriam Valdin, Tessa Vidal, Jérémy Viel, Alain Zamparutti.

WILDLIFE CONSULTANTS
Pierre Cadeac, Benoît Charrier, Christiane D'Hotel, Laurent and Michel Flaesch, Thomas Garrido, Christophe Guillard, Boris Juliot, Paul Lefranc, Tyler Nelson (United States), Jean-Philippe Varin.

MOUNTAIN GUIDES
Éric Alexandre, Denis Ducroz, Bernard Terraz, Marc Ziegler, Oscar Guineo (Chile).

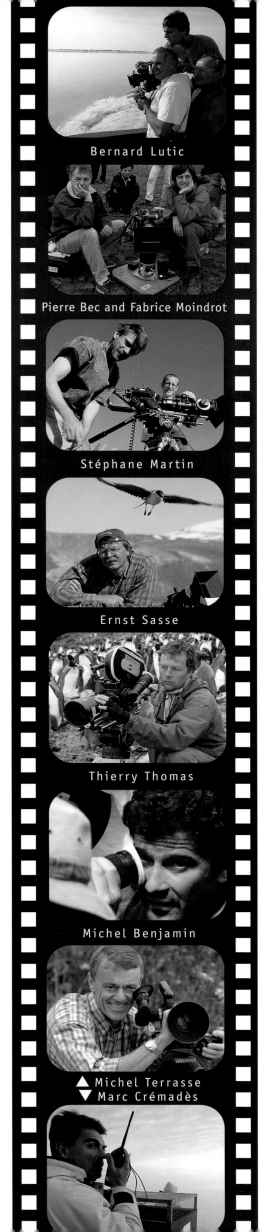

Bernard Lutic

Pierre Bec and Fabrice Moindrot

Stéphane Martin

Ernst Sasse

Thierry Thomas

Michel Benjamin

▲ Michel Terrasse
▼ Marc Crémadès

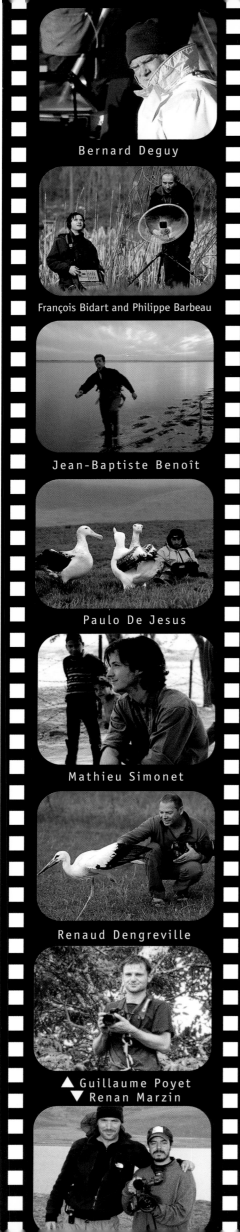

Bernard Deguy

François Bidart and Philippe Barbeau

Jean-Baptiste Benoît

Paulo De Jesus

Mathieu Simonet

Renaud Dengreville

▲ Guillaume Poyet
▼ Renan Marzin

Acknowledgments

EUROPE

FRANCE

The city of Paris; Police Headquarters Paris; General Director of Civil Aviation.

The National Office of Historic Monuments; Château du Haut-Koenigsbourg.

The town hall of Oloron-Sainte-Marie; the Oloron-Sainte-Marie Flying Club.

Paul-Sabatier University; the Pic du Midi Observatory; the Haute-Provence Observatory; the National Center for Scientific Research.

The Prefecture of Finistère; the French navy and the ships *La Loire* and *Latouche-Treville*; Brest Maritime Prefecture; the Directorate General of Maritime Customs and Excise for the Atlantic, Denis Ribaut; the Regional Crop Pest Control Service, Bruno Hamonet; Sesidec, Control of Animals Detrimental to Crops, Philippe Terrieux; SNSM (National Marine Salvage Society); the launch at Lézardrieux, J.-M. Touja; the town halls of Goulven, Kerlouan, Lampaul-Ploudalmezeau, and Saint-Pabu.

Admiral Delaunay, Commandant Lajous, SIRPA Mer, Pierre Dubrulle; SIRPA Air and La Patrouille de France, Commandant Girbes.

The Prefecture of La Manche; the Regional Council of Basse-Normandie: René Garrec, Nicole Ameline, Mr. Pommel; the General Council of Calvados, Anne d'Ornano, Anne-Marie Chuet-Deschamps; the Mare de Vauville Nature Reserve: Thierry Desmarets; the Calvados Hunters' Federation; the Saint-Samsom Nature Reserve; the Port of Le Havre; EDF/GDF Calvados, J.-P. Dalmar; the town halls of Deauville, Houlgate, Mont-Saint-Michel, and Cléville; the Normandy Ornithological Group, Gérard Debout; DSV (Department of Veterinary Services) in Calvados, Mr. Gautier-Chevreux, Ms. Blot & Ms. Steiner.

The Military Delegation for Eure-et-Loir; the Northwest Army Region; Domaine de Chambord, Mr. Hely at la Roche Aymon.

The Regional Council of Franche-Comté, Chantal Fischer, mayor of the commune of Montbenoît, Frédéric Bourdin; Champagnole-Crotenay Airfield.

The Regional Council of Languedoc-Roussillon, Jacques Blanc, Bertrand Bayle, and Dominique Poitevin; the General Council of Lozère, Jean-Paul Pottier; the town halls of Nasbinals and Malbouzon.

The General Council of Aveyron, Jean Puech, Alain Cardron, Véronique Bastide, and Claudine Veber; the town hall of Rodelle.

The Port of Dunkerque.

Brière Regional Park, Jacques Hedin; the Commission of la Grande-Brière Mottière, Félix Perraud; the Loire-Atlantique Maritime Navigation Service.

The La Tour du Valat Biological Study Station,

Jean-Paul Taris; SNPN (The National Society for the Protection of Nature); the National Reserve of Camargue.

The La Rochelle Maritime Museum, Patrick Schnepp.

The National Forests Office, Crest-Voland (Savoie), Jean-Baptiste Malinverno; the Deux Savoies Grouse Synergetic Interest Group; the town hall of Vallorcine (Haute-Savoie).

The French Austral and Antarctic Territories; the French Institute for Polar Research and Technology; the Préfecture of Guyana; the French Armed Forces High Command in Guyana, General Le Bourdonnec; the commandant of Saint-Laurent and Mana Gendarmerie Squadron, Captain Delgrande; Groupement Maroni 9th Rima, Lieutenant Colonel Secretin; chief of the Antilles/Guyana Group of the ONCFS (National Office of Hunting and Wild Animals), Éric Hansen; the curator of the Mana Nature Reserve, Xavier Desbois; Kwata, Ingrund Vogel.

Hafid Ba-Mohammed, Jacques Perrin-Thales, François Bel, Jacques Ségala (Euro RSCG), Dimitri Voevodsky, Pierre Fyot, and René.

GERMANY

Institut für Vogelforschung, Wilhelmshaven, Michael Exo, Franz Bairlein, Christiane Ketzenberg; Günter Nowald, director of the Center for Information on Cranes; Dr. Berndt-Ulrich Meyburg, director of the World Birds of Prey Group; Euronatur, Lutz Ribbe.

SPAIN

The General Directorate of the Environment of the Government of Navarre, Jaime Gaztelu, Fernando Portillo. The Regional Council of Extremadura; the Directorate of Agriculture and the Environment, Juan Pablo Prieto, Cipriano Hurtado, Santiago Corchette, Joaquin Gasset, Marquis de Gorbea; the Regional Council of Aragon; the Regional Council for the Environment of Huesca, Manuel Alcantara; the dovecotes of Extlar; the High Council for Scientific Research, Dr. Juan C. Alonso; the National Museum of Natural Sciences, Madrid; the Villafafila Bird Reserve, Ana Martinez, Andres Roman Vaquero, Mariano Rodriguez, Jesus Palacios, and Joaquin Araujo.

GREENLAND

Kalaallit Nunaat Nature Tourism, Silverio Scivoli.

THE NETHERLANDS

The Rotterdam Zoo, the Royal Rotterdam Zoological and Botanical Gardens, Stuster Koos.

ICELAND

Sigurdur A. Thraïnsson; the Nature Conservation Agency, Arni Braganson; the Ministry of the Environment, Sigmundur Einarsson, Ingimar Sigurdsson; the Ministry of Agriculture, the chief veterinary surgeon, Halldor Runolfsson, Jarle Reiersen; Baldur Kafnasson, Einar O. Thorleifsson.

ITALY

The National Institute for Wildlife Biology, Alessandro Ghi, Fernando Spina.

POLAND

The Ministry of the Environment, Janusz Radziejowski.

RUSSIA

The All-Russian Research Institute for Nature Protection, Alexander G. Sorokin; the Ministry of Protection of the Environment and Natural Resources of the Russian Federation and Anastassia Shilina, director of the Oka Biosphere Reserve, Dr. Yuri M. Markin. In memory of Vladimir Pachenko.

SWITZERLAND

The Federal Department of the Environment, Transport, Energy, and Communication; the Federal Office of Civil Aviation, Maurice Lenenberger.

NORTH AMERICA

CANADA

The Canadian Wildlife Service, Jean-Yves Charrette, Vicky Johnston, Lise Dussault.

Quebec: the Department of Biology and the Nordic Studies Center, Laval University; Professeur Gilles Gauthier; the Cap Tourmente National Wildlife Area, Serge Labonte; Forillon National Park, Jean-Guy Chavarie; Bonaventure Island and Percé Rock Park, Chantal Bourget; Professor Jacques Larochelle; Quentin Van Ginhoven; Martin Dignard.

Nunavut: the Nunavut Impact Review Board; the Nunavut Research Institute, Mary Ellen Thomas; the Continental Polar Plateau Study; the Canadian National Parks; Siqmillik National Park, Bylot Island.

UNITED STATES

The United States Department of Agriculture, Sara Kaman, Dr. Kay W. Wheeler; the Federal Aviation Administration, Al Pereira, Ted Disantis; the U.S. Fish and Wildlife Service; the Bureau of Citizenship and Immigration Services.

Bill Lischman; Partners Animal Institute.

Leighton Pakes; Milt's Menagerie, Valley Springs, Milton Scholten.

New York: Mayor's Office of Film, Theater, and Broadcasting, City of New York; the Adirondack Regional Tourism Council, the office of the project director of Liberty State Park, Ann Melious and Brenda McKinley; United States Ultralight Association, Tom Gunnarson; New York LaGuardia International Airport air-traffic control tower, Michael J. Sammartino, Cecilia Castro.

Oregon: the Klamath Falls Chamber of Commerce, Brian Baxter.

Utah: the Utah Film Commission; the Utah Travel Council, Tracie Cayford, Ken Kraus; biologist Victoria Roy.

Idaho: the Idaho State Department of Fish and Game, Jack Connelly.

Colorado: the Colorado Division of Wildlife Research Center, Clait E. Braun.

Nebraska: the Nebraska Film Office, Laurie Richards; the International Cranes Foundation, George Archibald; the Rowe Sanctuary, Paul Tebbel, Lilian Annette.

Alaska: the Alaska Film Office, Mary Pignalberi and Julie Ford; the U.S. Forest Service, Mary Anne Bishop; the Alaska Chilkat Bald Eagle Preserve, Bill Zack; the Pacific Northwest Research Station; the Cooper River Delta Institute.

Arizona: the Arizona Film Commission; the Tucson Film Office Commission, Peter and Chelly; the U.S. Bureau of Land Management; the Glen Canyon National Recreation Area (Lake Powell), Lynn Picard, park ranger; the Tonto National Forest (Lake Roosevelt).

Montana: the Montana Film Office, Sten Iversen; the Freezout Lake Wildlife Management Area, Mark Schlepp; Ralph and Kathleen Waldt.

Colorado: the Navajo Nation Film Office (Monument Valley), Kee Long; the Hualapai Indian Reservation (Grand Canyon), Sandra Yellowhawk and Michelle.

North Dakota: Department of Zoology, North Dakota University, Gary L. Nuechterlein.

SOUTH AMERICA

ARGENTINA

Nahuel Huapi National Park; Don Andres Domingo, La Buitrera; the La Plata Ornithological Association, Santiago Krapovickas. Juan José Arenas and Don Andres Domingo; Luis Noberto Jacome and Sergio Lambertucci.

CHILE

Torres Del Paine National Park.

FALKLAND ISLANDS

The Falkland Islands government, Donald A. Lamont, governor; New Island Parks.

PERU

Inrena (the National Institute of National Resources); Rainforest Expeditions, Luis Zapater, Guillermo Knell; the Tambopata Research Center.

AFRICA

KENYA

African Latitude, Michel and Robyn Laplace-Toulouse.

LIBYA

Mohamed Ahmed Alaswad, Libyan ambassador to UNESCO; the minister of tourism, Mr. Boukhari; Dr. Fathi M. Elmusrati and Dr. Aymen Seif Ennasir; Ministry of Transport, Bachir Emar; Libyan Airlines, Adem Abdoulmagid; the Libyan Veterinary Service and the Tripoli Zoo, Dr. Jumma Haluai; Zawia Travels.

MALI

The Ministry of Culture and Tourism; the National Center for Cinema Production, Youssouf Coulibaly; Ladji Diakite, Sidi Becaye Traore.

MAURITANIA

The Ministry of Information, Abdallahi ould Loudaa; the Ministry of Communication, Rachid Ould Saleh; the Civil Aviation Directorate, Boirick Ould Gharve; the Arguin Banks National Park, Mohamed Ould Bouceif; FIBA (the National Arguin Banks Foundation), Luc Hoffman, and scientific adviser Jean Worms.

SENEGAL

Djoudj: the Ministry of Culture and Communication, Mamadou Diop de Croix; the Ministry of the Environment, M. Lamine Ba; the Ministry of Public Works and Transport; the National Parks Directorate of Senegal; Aïda Ba; the Djoudj National Bird Sanctuary, Commandant Souleye Ndiaye, Commandant Demba Mamadou Ba, Colonel Sara Diouf, Captain Ibrahim Diop, Lieutenant Sidibe, Indega Bindia.

ASIA

SOUTH KOREA

The Department of Natural Sciences of Kyungnam University, Prof. Kyu-Hwang Hahm; the Korean Ornithological Institute; Kyung Hee University, Prof. Seong-Hwan Pae.

INDIA

The Bharatpur Bird Sanctuary, Shruti Sharma, Bholu Khjan.

JAPAN

Hokkaido: the Wild Bird Society of Japan, Turui-Itoh Tanchou Sanctuary, John O. Albersten; East-Hokkaido, National Park and Wildlife, Yulia Momose; the Akkeshi Marine Biological Station, Testuro Kaji, Osamu Harada.

NEPAL

His Majesty's Government of Nepal; the Ministry of Culture; the Department of National Parks and Wildlife Conservation; Summit Trekking, Kit Spencer and Amar, the inhabitants of the town of Kagbeni.

PHILIPPINES

Birds International Inc., Antonio M. De Dios.

VIETNAM

The Ministry of Culture and Information; Studio No. 1 Hanoi, the Center for the Study of the Environment and Natural Resources.

SCIENTIFIC PARTNERS

The National Museum of Natural History, Paris Zoological Park, Bois de Vincennes, Prof. Jean-Jacques Petter, Jean-Louis Deniaud; the Jardin des Plantes Menagerie, Jacques Rigoulet, the Laboratory of Parasite Biology, Annie Petter and Odile Bain; the Mammals and Birds Library, Évelyne Bremond and Jean-Marc Bremond; the National Museum of Natural History, Cleres, Dr. Ing A. Hennache; the Museum of Natural History of Dijon, Dominique Geoffroy, the Arctic Ecology Research Group, Brigitte Sabard, Olivier Gilg; the Natural History Museum of Lille, Yves Gometou, the CNRS laboratory in Chize, CEBC, M. Henry Weimerkirsch; the CNRS laboratory in Strasbourg, Yvon Le Maho; the Analysis Laboratory of the Departments, Dr. J. P. Buffereau; the Lorraine Conservatory of Natural Sites, Alain Salvi; Villars-les-Dombes Wildlife Park, Bernard Fulcheri, Éric Bureau, Jean-François Lefèvre; Marquanterre Park, Nicolas Durand; the Bird Garden, Mr. Liauzu; Saint-Martin-de-la-Plaine Park, Mr. Thivillon and Mr. Boussekey; Doué-la-Fontaine Zoo, Pierre Gay, Brice Lefaux; the Orangerie Zoo, Claude Rink; Tregomeur Zoological Gardens, Mr. Arnoux; Cleres Wildlife Park, Jean Delacour, Alain Hennache, Michel St. James; Paugres Safari Park, Christelle Vitaud; Ker Anas Bird Park, Philippe Rambaud; Branfere Park, Fondation de France, Y. Philippot; Dr. Éric Plouzeau; Jean-Philippe Varin, the Jacana Wildlife Studio; Michel Beaurin; Mr. Gibouin; Michel Ledoux; Dr. Henri Quinque; Mr. Terran; Mr. Wilman, Top Duck, and Michel Beaurin.

FRENCH MINISTRIES AND OFFICIAL ORGANIZATIONS

The General Secretary of the Élysée, Dominique de Villepin; the Ministry of Foreign Affaires: the Office of African and North African Affairs and its director, J. D. Roisin; the Latin American Office and its director, J. M. Laforêt; the Asia and Oceania Office; technical adviser for Vietnam, J. B. Lesecq; the Ministry of Public Works, Transport, and Housing, minister Jean-Claude Gayssot and chargé de mission Bernard Vasseur; the office of the Junior Minister for Foreign Aid, minister Charles Josselin.
The European Commission, the Environment Office, Delphine Malard, administrator; the Consulate of the United States, Paris, Anne Syret, chief visa officer.

THE EMBASSIES

We would like to thank all the French embassies abroad and foreign embassies in France, and, particularly, the French embassy in Iceland and Ambassador Rob Cantoni; the French embassy in the United States and Véronique Godard of the Audiovisual Service; the French embassy in Peru; the Peruvian embassy and consulate in France; the French embassy in Mauritania, Ambassador Jean-Paul Taix and First Secretary Martin Juillard; the French embassy in Libya, Ambassador Josette Dalland, Cultural Attaché Benoît Deslandes; the French embassy in Nepal, Ambassador Ambrosini; the United States embassy in Nepal, Ambassador Ralph Franck; the French embassy in Japan, Moriyuki Motono; the French embassy in Vietnam, Ambassador Serge Degallaix; the military attaché, Colonel Protar, the audiovisual attaché, Olivier Delpoux; the French embassy in Sweden, Ambassador Patrick Imhaus.

THE SPONSORS

Bateaux Jeanneau, Roland Fardeau, and Pierre André; Citroën, Yves Boutin and Isabelle Seyller; Établissements Guy Cotten, Guy Cotten and Lionel Guiban; Kodak; Matra-Auto, Enzo Garavelloni; Nikon; Objectif Bastille; Le Vieux Campeur; Yamaha Motor France, Denis Ricard and Patrick Jacquin; Zodiac International, Pierre Barbleu.

FRENCH COPRODUCERS

Les Productions de La Guéville; Bac Films; France 2 Cinéma; France 3 Cinéma.

FOREIGN COPRODUCERS

Pandora Films/Germany, Karl Baumgartner, Reinhard Brundig; Les Productions JMH/Switzerland, Jean-Marc Henchoz; Wanda Vision/Spain, José-Maria Morales; Eyescreen/Italy, Andrea Occhipinti.

PARTNERS

Canal+; EDF, CEO François Roussely; Crédit Agricole, public-relations director Didier Blaque-Belair and Pierre Moulies; CFACE, François David; Lufthansa German Airlines, public relations director Lutz Lammerhold, Claudia Ungeheuer and Christoph Potting (Ahrens & Behrent); Primagaz, Alain Rousseau; Airbus; OBC, Didier Kunstlinger; Cofiloisirs, Denis Offroy and Nicole Hyde; the National Center for Cinematography; Procirep; the GAN Foundation for Cinema; the European Commission (General Directorate of the Environment); the Eurimages Fund of the Council of Europe; the National Museum of Natural History, its administrator, M. Moreno, its audiovisual director, J. P. Baux; the World Wildlife Foundation; the League for the Protection of Birds.

FOREIGN SALES:

Jacques E. Strauss, Président films.

And particular thanks to Gérard Vienne, at whose side it all began with *The Monkey People*.

ADDITIONAL PHOTOGRAPHIC HELP

Patrick Chauvel, Luc Coutelle, Philippe Garguil, Aude Mesnil, Christophe Pottier, and Michel Terrasse.

All photographs were taken with the Nikon F-100 and the Nikon range of lenses.

Photographic Credits

Laurent Charbonnier: 37, 268e.

Patrick Chauvel: 235 top, 246cg.

Jacques Cluzaud: 80–81.

Luc Coutelle: 120–121, 146–147, 207 left, 254 top.

Marc Cremades: 226a, 251 bottom, 268h.

Renaud Dengreville: 18–19, 38, 58–59, 68–69, 79, 95, 100–101, 116–117, 118–119, 140, 142–143, 150–151, 153, 158–159, 160–161, 164–165, 199, 201, 202–203, 216 left, 226e, 229 top, 232c, 241 top, 246af, 249a, 251 top, 268b, spread 4c.

Jean-Patrick Deya: 266h.

Christiane D'Hôtel: 246b.

Stéphane Durand: 268c.

Philippe Garguil: 32–33, 35, 186, 187, 268g.

Frédéric Labrouche: 227 top.

Michel Laplace-Toulouse: 222 bottom.

Toinette Laquière: 239 top.

Paola Luttringer: 232b.

Renan Marzin: 41, 48a, 104–105, 125, 131, 174–175, 222 top, 232d, 240bd, 247, 250, 253, 267acd.

Aude Mesnil: spread 2b.

Johann Mousseau: 267f.

Christophe Pottier: 24–25, 54–55, 56–57, 172–173, 176–177, 179, 184, 220, 246d, 266f, 267e, 268d.

Guillaume Poyet: 36, 47, 48bc, 49, 51, 64–65, 83, 84–85, 87, 91, 99, 106–107, 130, 132–133, 134, 135, 168–169, 170–171, 178, 180–181, 182–183, 185, 188, 189, 190, 191, 192, 193, 197, 206 right, 210 left, 212 right, 213, 217 middle, 226g, 257 bottom, 259, 260–261, 263, 265, spread 1, spread 3abc.

Mathieu Simonet: flyleaf, 2, 4–5, 6–7, 8–9, 10–11, 12–13, 15, 16–17, 20–21, 22–23, 26–27, 28–29, 30–31, 39, 42, 43, 45, 46, 50, 52–53, 60–61, 62–63, 66–67, 70–71, 72–73, 74–75, 76–77, 78, 88–89, 93, 96, 97, 102–103, 108–109, 110–111, 112–113, 114–115, 122–123, 126, 128ab, 129, 136–137, 138–139, 141, 144–145, 148–149, 152, 154–155, 156–157, 162–163, 166–167, 194–195, 205, 206 left, 207 right, 210 middle, 210 right, 216 right, 217 left, 217 right, 222 middle, 223, 224–225, 226bcdf, 227 bottom, 229 bottom, 231, 232aefg, 233, 235 bottom, 236, 239 bottom, 240ace, 241 bottom, 243, 245, 246e, 249bcd, 254 bottom, 255, 257 top, 262, 266abcdeg, 267bh, 268af, 270, 271, spread 2acd, spread 3d, spread 4abd.

Michel Terrasse: 82, 212 left, 267g.

Watercolors by Damien Chavanat: 234, 248.

Cartography: Édigraphie, Rouen.

Drawings from the storyboard by Olivier Cheres: 230, 238, 242, 244.

Originally published in France in 2001 by Éditions du Seuil, under the title
Le Peuple migrateur
Copyright © 2001 by Éditions du Seuil.
Original ISBN 2-02-050566-5.
English translation copyright © 2003 by Éditions du Seuil.

Library of Congress Cataloging-in-Publication Data available.
English translation by David Wharry
ISBN 2-02-061292-5

Book design by Valérie Gautier.
English type design by Flux.
Typeset in Bauer Bodoni, Garamond and Officina.
Manufactured in France.

Distributed in Canada by Raincoast Books
9050 Shaughnessy Street, Vancouver, British Columbia V6P 6E5

10 9 8 7 6 5 4 3 2

Chronicle Books LLC
85 Second Street, San Francisco, California 94105
www.chroniclebooks.com